Lecture Notes in Mathematics 1504

J. Cheeger M. Gromov C. Okonek P. Pansu

Geometric Topology: Recent Developments

Lectures given on the 1st Session of the
Centro Internazionale Matematico Estivo (C.I.M.E.)
held at Montecatini Terme, Italy, June 4-12, 1990

Editors: P. de Bartolomeis, F. Tricerri

Springer-Verlag
Berlin Heidelberg New York
London Paris Tokyo
Hong Kong Barcelona
Budapest

Authors

Jeff Cheeger
Courant Institute, NYU
251 Mercer Street
New York, NY 10012, USA

Mikhail Gromov
IHES
35 rue de Chartres
91440 Bures-sur-Yvette, France

Christian Okonek
Mathematisches Institut der
Universität Bonn
Wegelerstrasse 10
5300 Bonn, Germany

Pierre Pansu
Centre de Mathématiques
Ecole Polytechnique
91128 Palaiseau, France

Editors

Paolo De Bartolomeis
Dipartimento di Matematica
Applicata G. Sansone
Via di S. Marta, 3
50139 Firenze, Italy

Franco Tricerri
Dipartimento di Matematica U.DINI
Viale Morgagni 67/A
50134 Firenze, Italy

Mathematics Subject Classification (1991): 53C20, 53C32, 22E40, 14J60, 32L07, 57R55

ISBN 3-540-55017-8 Springer-Verlag Berlin Heidelberg New York
ISBN 0-387-55017-8 Springer-Verlag New York Berlin Heidelberg

Typesetting: Camera ready by author
Printing and binding: Druckhaus Beltz, Hemsbach/Bergstr.
46/3140-543210 - Printed on acid-free paper

Foreword

The present volume contains the text of three series of lectures given in Montecatini for the period June 4-June 10, 1990, during the C.I.M.E. session "Recent Developments in Geometric Topology and Related Topics".

Geometric Topology can be defined to be the investigation of global properties of a further structure (e. g. differentiable, Riemannian, complex, algebraic etc...) assigned to a topological manifold.
As a result of numerous recent outstanding achievements, which are as complex as they are deep, and always involve a dramatic spectrum of tools and techniques originating from a wide range of domains, Geometric Topology appears nowadays as one of the most fascinating and promising fields of contemporary mathematics.

Our main goal in organizing the session was to gather a distinguished group of mathematicians to update the subject and to give a glimpse on possible future developments.
We can proudly affirm that the lecturers did a superb job.
For an idea of how rich and interesting was the subject-matter that they presented, it is enough to give a brief description of the three main topics.

[1] The geometry and the rigidity of discrete subgroups in Lie groups especially in the case of lattices in semi-simple groups.Two main streams of approaches are considered:
i) the geometry and the dynamics of the action of discrete subgroups on the ideal boundary of the ambient group;
ii) the theory of local and infinitesimal deformations of discrete subgroups via elliptic P.D.E. and Bochner type integro-differential inequalities.
The basics of these two methods are fully described and more advanced materials are covered.
[2] The study of the critical points of the distance function and its application to the understanding of the topology of Riemannian manifolds.
Moving from Toponogov's celebrated theorem and from a complete description of the techniques of critical points of distance function, three basic results in global differential geometry are discussed:
i) the Grove-Petersen theorem of the finiteness of homotopy types of manifolds admitting metrics with bounds on diameter, volume and curvature;
ii) Gromov's bound on the Betti numbers in terms of curvature and diameter;
iii) the Abresch-Gromoll theorem on finiteness of topological type, for manifolds with nonnegative Ricci curvature, curvature bounded below and slow diameter growth.

[3] The theory of moduli space of instantons as a tool for studying the geometry of low-dimensional manifolds.

As main topics, we can quote:

i) the correspondence between instantons over algebraic surfaces and stable algebraic vector bundles, with the investigation of the relations between the geometry of an algebraic surface and the differential topology of its underlying 4-manifold;

ii) the existence of infinitely many exotic C^∞-structures on some topological 4-manifolds;

iii) the theory of the decomposition of 4-manifolds along homology 3-spheres.

Finally, it is worthwhile adding that the texts of the present volume capture completely the spirit and the atmosphere of a very successful event.

Paolo de Bartolomeis

Franco Tricerri

TABLE OF CONTENTS

Foreword V

J. CHEEGER, Critical Points of Distance Functions and Applications
to Geometry 1

M. GROMOV, P. PANSU, Rigidity of Lattices: An Introduction 39

CHR. OKONEK, Instanton Invariants and Algebraic Surfaces 138

List of participants 187

Critical Points of Distance Functions and Applications to Geometry

Jeff Cheeger

0. Introduction
1. Critical points of distance functions
2. Toponogov's theorem; first applications
3. Background on finiteness theorems
4. Homotopy Finiteness
   Appendix. Some volume estimates
5. Betti numbers and rank
   Appendix: The generalized Mayer-Vietoris estimate
6. Rank, curvature and diameter
7. Ricci curvature, volume and the Laplacian
   Appendix. The maximum principle
8. Ricci curvature, diameter growth and finiteness of topological type.
   Appendix. Nonnegative Ricci curvature outside a compact set.

0. Introduction

These lecture notes were written for a course given at the C.I.M.E. session "Recent developments in geometric topology and related topics", June 4-12, 1990, at Montecatini Terme. Their aim is to expose three basic results in riemannian geometry, the proofs of which rely on the technique of "critical points of distance functions" used in conjunction with Toponogov's theorem on geodesic triangles. This method was pioneered by Grove and Shiohama, [GrS].

Specifically, we discuss

i) the Grove-Petersen theorem of the finiteness of homotopy types of manifolds admitting metrics with suitable bounds on diameter, volume and curvature; [GrP],

ii) Gromov's bound on the Betti numbers in terms of curvature and diameter; [G],

iii) the Abresch-Gromoll theorem on finiteness of topological type, for manifolds with nonnegative Ricci curvature, curvature bounded below and slow diameter growth; [AGl].

The first two of these theorems are stated in § 3 and proved in § 4 and §§ 5-6, respectively. The third is stated and proved in § 8.

The reader is assumed to have a background in riemannian geometry at least the rough equivalent of the first six chapters of [CE], and to be familiar with basic algebraic topology. For completeness however, the statement of Toponogov's theorem is recalled in § 2. Additional material on finiteness theorems and on Ricci curvature is provied in § 3 and § 7.

1. Critical Points of Distance Functions.

Let M^n be a complete riemannian manifold. We will assume that all geodesics are parametrized by arc length. For $p \in M^n$, we denote the distance from x to p by $\overline{x,p}$ and put

$$\rho_p(x) := \overline{x,p}$$

Note that $\rho_p(x)$ is smooth on $M \setminus \{p \cup C_p\}$, where C_p, the *cut locus* of p, is a closed nowhere dense set of measure zero.

Grove and Shiohama made the fundamental observation that there is a meaningful definition of "critical point" for such distance functions, such that in the absence of critical points, the Isotopy Lemma of Morse Theory holds. They also observed that in the presence of a lower curvature bound, Toponogov's theorem can be used to derive geometric information, from the *existence* of critical points. They used these ideas to give a short proof of a generalized Sphere Theorem, see Theorem 2.5. Other important applications are discussed in subsequent sections.

Remark 1.1. If the *sectional curvature* satisfies $K_M \le K$ (for $K \ge 0$) and q is a critical point of ρ_p with $\rho_p(q) \le \frac{\pi}{2\sqrt{K}}$, then there is also a reasonable notion of *index* which predicts the change in the *topology* when crossing a critical level. But so far, this fact has not had strong applications.

Definition 1.2. The point q $(\neq p)$ is a *critical point* of ρ_p if for all v in the tangent space, M_q, there is a minimal geodesic, γ, from q to p, making an angle, $\measuredangle\,(v, \gamma'(0)) \le \frac{\pi}{2}$, with $\gamma'(0)$. Also, p is a critical point of ρ_p.

From now on we just say that q is a critical point of p.

Remark 1.3. If $q \neq p$ is a critical point of p, then $q \in C_p$. If q is *not* critical, the collection of tangent vectors to all geodesics, γ, as above, lies in some *open* half space in M_q. Thus, there exists $w \in M_q$, such that $\measuredangle\,(w, \gamma'(0)) < \frac{\pi}{2}$, for all minimal γ from p to q.

Put $B_r(p) = \{x \mid \overline{x,p} < r\}$.

Isotopy Lemma 1.4. *If $r_1 < r_2 \le \infty$, and if $\overline{B_{r_2}(p)} \setminus B_{r_1}(p)$ is free of critical points of ρ_p, then this region is homeomorphic to $\partial B_{r_1}(p) \times [r_1, r_2]$. Moreover, $\partial B_{r_1}(p)$ is a topological submanifold (with empty boundary).*

Proof: If x is noncritical, then there exists $w \in M_x$ with $\measuredangle\,(\gamma'(0), w) < \frac{\pi}{2}$, for all minimal γ from x to p. By continuity, there exists an extension of w to a vector field, W_x, on a neighborhood, U_x, of x, such that if $y \in U_x$ and σ is minimal from y to p, then $\measuredangle\,(\sigma'(0), W_x(y)) < \frac{\pi}{2}$. Take a finite open cover of $\overline{B}_{r_2}(\rho) \setminus B_{r_1}(\rho)$, by sets, U_{x_i}, locally finite if $r_2 = \infty$, and a smooth partition of unity, $\sum \phi_i \equiv 1$, subordinate to it. Put $W = \sum \phi_i W_{x_i}$. Clearly, W is nonvanishing. For each integral curve ψ of W, the *first variation formula* gives

$$\rho_p(\psi(t_2)) - \rho_p(\psi(t_1)) \leq (t_1 - t_2)\cos(\frac{\pi}{2} - \epsilon)\,,$$

for some small ϵ. This holds on compact subsets if $r_2 = \infty$. The first statement easily follows.

To see that $\partial B_{r_1}(p)$ is a submanifold, let $q \in \partial B_{r_1}(p)$, σ a minimal geodesic from q to p, and V a small piece of the totally geodesic hypersurface at q, normal to σ. Then for $z \in V$, sufficiently close to q, each integral curve, ψ, of W through z intersects $\partial B_{r_1}(p)$ in exactly one point, $z' \in \partial B_{r_1}(p)$ (ψ extends on both sides of V). It is easy to check that the map, $z \to z'$, provides a local chart for $\partial B_{r_1}(p)$ at q.

Example 1.5. M compact and q a farthest point from p implies that q is a critical point of ρ_p, obviously, the topology changes when we pass q. This observation was made by Berger, well in advance of the formal definition of "critical point"; [Be].

Example 1.6. If γ is a geodesic loop of length ℓ and if $\gamma \mid [0, \frac{\ell}{2}]$ and $\gamma \mid [\frac{\ell}{2}, \ell]$ are minimal, then $\gamma(\frac{\ell}{2})$ is a critical point of $\gamma(0)$. In particular, if q is a closest point, to p on C_p, and q is not conjugate to p along some minimal geodesic then q is a critical point of p; see Chapter 5 of[CE]. Thus, if p, q realize the shortest distance from a point to its cut locus in M^n, and are not conjugate along any minimal γ, then p and q are *mutually critical.*

Example 1.7. On a flat torus with fundamental domain a rectangle, the barycenters of the sides and the corners project to the three critical points of p, other than p itself.

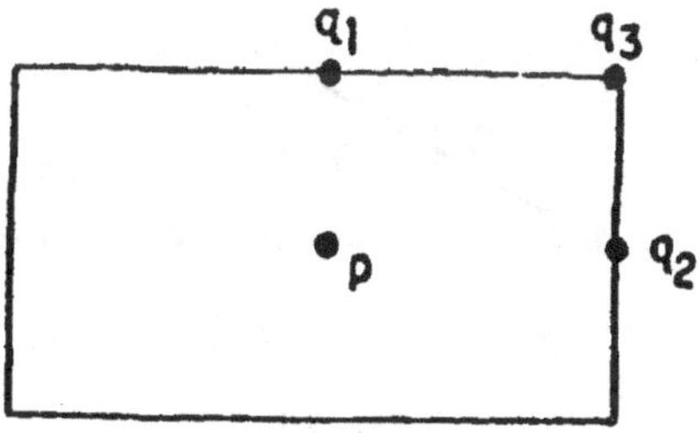

Fig. 1.1

Example 1.8. A conjugate point need not be critical. Here is a concrete example. Write the standard metric on S^2 in the form $g = dr^2 + \sin^2 r\, d\theta^2$, where $0 \leq r \leq \pi$, $0 \leq \theta \leq 2\pi$. Let $f(r, \theta)$ be a smooth function, periodic in θ, such that

i) $f(r,\theta) \equiv 1$, for all (r, θ) satisfying any of the following conditions:

$$0 \leq r \leq \frac{\pi}{4}\,, \qquad \frac{3}{4}\pi \leq r \leq \pi\,,$$
$$\pi - \epsilon \leq \theta \leq \pi + \epsilon\,.$$

Here we require $\epsilon < \pi/4$.

ii) $f > 1$ elsewhere.

The metric $g' = f\,dr^2 + \sin^2 r\,d\theta^2$ satisfies $g' \geq g$. In fact, if the intersection of a curve, c, with the region, $\pi/4 < r < 3\pi/4$, is not contained in the region $\pi - \epsilon \leq \theta \leq \pi + \epsilon$, then its length with respect to g' is strictly longer than with respect to g. It follows that for the metric g', the only minimal geodesics connecting the "south pole" ($\theta = \theta$) to the "north pole", ($\theta = \pi$) are the curves $c(t) = (t, \theta_0)$, $\pi - \epsilon \leq \theta_0 \leq \pi + \epsilon$. Since $2\epsilon < \pi$, it follows that the north and south poles are mutually *conjugate*, but mutually *noncritical.*

We are indebted to D. Gromoll for helpful discussions concerning this example.

Remark 1.9. The *criticality radius*, r_p, is, by definition, the largest r such that $B_r(p)$ is free of critical points. By the Isotopy Lemma 1.4, $B_{r_p}(p)$ is homeomorphic to a standard open ball, since it is homeomorphic to an arbitrarily small open ball with center p.

2. Toponogov's Theorem; first applications.

Denote the length of γ by $L[\gamma]$.

By definition, a *geodesic triangle* consists of three geodesic segments, γ_i, of length $L[\gamma_i] = \ell_i$, which satisfy

$$\gamma_i(\ell_i) = \gamma_{i+1}(0) \bmod 3 \quad (i = 0, 1, 2)\ .$$

The *angle* at a corner, say $\gamma_0(0)$, is by definition, $\measuredangle(-\gamma_2'(\ell_2), \gamma_0'(0))$. The angle opposite γ_i will be denoted by α_i.

A pair of sides e.g. γ_2, γ_0 are said to determine a *hinge*.

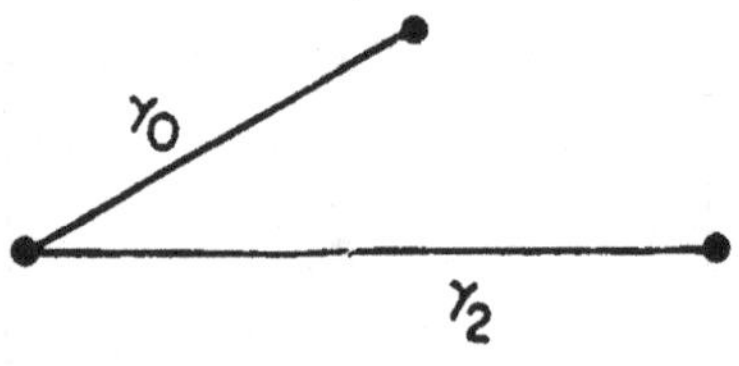

Fig. 2.1

Let M_H^n denote the n-dimensional, simply connected space of curvature $\equiv H$ (i.e. hyperbolic space, Euclidean space, or a sphere).

Toponogov's theorem has two statements. These are equivalent in the sense that either one can easily be obtained from the other.

Theorem 2.1 (Toponogov). *Let M^n be complete with curvature $K_M \geq H$.*

A) *Let $\{\gamma_0, \gamma_1, \gamma_2\}$ determine a triangle in M^n. Assume γ_1, γ_2 are minimal and $\ell_1 + \ell_2 \geq \ell_0$. If $H > 0$, assume $L[\gamma_0] \leq \frac{\pi}{\sqrt{H}}$. Then there is a triangle $\{\underline{\gamma}_0, \underline{\gamma}_1, \underline{\gamma}_2\}$ in M_H^2, with $L[\gamma_i] = L[\underline{\gamma}_i]$ and $\underline{\alpha}_1 \leq \alpha_1$, $\underline{\alpha}_2 \leq \alpha_2$.*

B) *Let* $\{\gamma_2, \gamma_0\}$ *determine a hinge in* M^n *with angle* α. *Assume* γ_2 *is minimal and if* $H > 0$, $L[\gamma_0] \leq \frac{\pi}{\sqrt{H}}$. *Let* $\{\underline{\gamma}_2, \underline{\gamma}_0\}$ *determine a hinge in* M_H^2 *with* $L[\gamma_i] = L[\underline{\gamma}_i]$, $i = 0, 2$, *and the same angle* α. *Then*

$$\overline{\gamma_2(0), \gamma_0(\ell_0)} \leq \overline{\underline{\gamma}_2(0), \gamma_0(\ell_0)}\,.$$

Proof: See [CE], Chapter 2.

Remark 2.2. In the applications of Toponogov which occur in the sequel, the following elementary fact is often used without explicit mention. Consider the collection of hinges, $\{\underline{\gamma}_0, \underline{\gamma}_2\}$ in M_H^2, with fixed side lengths, ℓ_0, ℓ_2 and variable angle α: $0 \leq \alpha \leq \pi$. *Then* $\overline{\underline{\gamma}_0(\ell_0), \underline{\gamma}_2(0)}$ *is a strictly increasing function of* α.

Remark 2.3. If the inequalities in A) or B) are all equalities, more can be said (see [CE]).

By using Toponogov's theorem we can derive geometric information from the existence of critical points.

Let the triangle, $\{\gamma_0, \gamma_1, \gamma_2\}$ satisfy the hypothesis of Toponogov's theorem, and assume $\gamma_0(\ell_0)$ is *critical* with respect to $\gamma_0(0)$. Then (as explained in detail in the applications), we can

i) bound from above the side length ℓ_2 (see Theorems 2.5, 4.2),

ii) bound from below, the excess, $\ell_0 + \ell_1 - \ell_2$ (see Proposition 8.5),

iii) bound from below, the angle α_1 (see Lemma 2.6, Corollaries 2.7, 2.9, 2.10 and 6.3).

Remark 2.4. It is important to realize that in order to obtain the preceding bounds, we do *not* assume $\alpha_2 \leq \pi/2$. The assumption that $\gamma_1(\ell_0)$ is *critical* with respect to $\gamma_0(0)$ implies that $\sphericalangle(-\tilde{\gamma}_0'(\ell_0), \gamma_1'(0)) \leq \pi/2$, for *some* minimal $\tilde{\gamma}_0$ from $\gamma_0(0)$ to $\gamma_0(\ell_0)$. This is all that we require.

Theorem 2.5 (Grove-Shiohama). *Let* M^n *be complete, with* $K_M \geq H$, *for some* $H > 0$. *If* M^n *has diameter*, $\mathrm{dia}(M^n) > \frac{\pi}{2\sqrt{H}}$, *then* M^n *is homeomorphic to the sphere*, S^n.

Proof: Let $p, q \in M^n$ be such that $\overline{p,q} = \mathrm{dia}(M^n)$; in particular, p and q are mutually critical (see Example 1.5).

Claim. There exists no $x \neq q, p$ which is critical with respect to p (the same holds for q).

Proof of Claim: Assume x is such a point. Let γ_2 be minimal from q to x. By assumption there exists γ_0, minimal from x to p, with

$$\alpha_1 = \sphericalangle(-\gamma_2'(\ell_2), \gamma_0'(0)) \leq \frac{\pi}{2}\,.$$

Similarly, since p and q are mutually critical, there exist *minimal* γ_1, $\tilde{\gamma}_1$ from p to q such that

$$\sphericalangle(-\gamma_0'(\ell_0), \gamma_1'(0)) \leq \frac{\pi}{2}$$

and

$$\sphericalangle(-\tilde{\gamma}_1'(\ell_1), \gamma_2'(0)) \leq \frac{\pi}{2}\,.$$

Note that $L[\gamma_1] = L[\tilde{\gamma}_1] = \overline{p,q} > \frac{\pi}{2\sqrt{H}}$.

Apply A) of Toponogov's theorem to both $\{\gamma_0, \gamma_1, \gamma_2\}$ and $\{\gamma_0, \tilde{\gamma}_1, \gamma_2\}$. Since a triangle in M_H^2 (the sphere) is determined up to congruence by its side lengths, we get a *unique* triangle, $\{\underline{\gamma}_0, \underline{\gamma}_1, \underline{\gamma}_2\}$, in M_H^2, *all* of whose angles are $\leq \pi/2$. By elementary spherical trigonometry, this implies that all sides have length $\leq \frac{\pi}{2\sqrt{H}}$, contradicting $\overline{p,q} > \frac{\pi}{2\sqrt{H}}$.

Given the claim, the proof is easily completed (compare the proof of Reeb's Theorem given in [M]).

The following observation and its corollaries (2.7, 2.10) are of great importance.

Lemma 2.6 (Gromov). *Let q_1 be critical with respect to p and let q_2 satisfy*

$$\overline{p, q_2} \geq \nu \overline{p, q_1} ,$$

for some $\nu > 1$. Let γ_1, γ_2 be minimal geodesics from p to q_1, q_2 respectively and put $\theta = \sphericalangle (\gamma_1'(0), \gamma_2'(0))$.

i) *If $K_M \geq 0$,*

$$\theta \geq \cos^{-1}(1/\nu) .$$

ii) *If $K_M \geq H$, $(H < 0)$ and $\overline{p, q_2} \leq d$, then*

$$\theta \geq \cos^{-1}\left(\frac{\tanh(\sqrt{-H}d/\nu)}{\tanh(\sqrt{-H}d)}\right) .$$

Proof: Put $\overline{p, q_1} = x$, $\overline{q_1, q_2} = y$, $\overline{p, q_2} = z$. Let σ be minimal from q_1 to q_2. Since q is critical for p, there exists τ, minimal from q to p with

$$\sphericalangle (\sigma'(0), \tau'(0)) \leq \frac{\pi}{2} .$$

i) Applying Toponogov's Theorem B) to the hinges $\{\sigma, \tau\}$ and $\{\gamma_1, \gamma_2\}$ gives

$$z^2 \leq x^2 + y^2 ,$$

$$y^2 \leq x^2 + z^2 - 2xz\cos\theta \quad \text{(law of cosines)}$$

Since $z \geq \nu \cdot x$, the conclusion easily follows.

ii) By scaling, we can assume $H = -1$. Replace the inequalities above by the following ones from hyperbolic trigonometry (see e.g. [Be])

$$\cosh z \leq \cosh x \cosh y ,$$

$$\cosh y \leq \cosh x \cosh z - \sinh x \sinh z \cos\theta .$$

Substituting the second of these into the first and simplifying gives

$$\theta \geq \cos^{-1}\left(\frac{\tanh x}{\tanh z}\right) ,$$

which suffices to complete the proof.

Corollary 2.7. *Let $q_1, \ldots, q_N$ be a sequence of critical points of p, with*

$$\overline{p, q_{i+1}} \geq \nu \, \overline{p, q_i} \quad (\nu > 1)$$

i) *If $K_{M^n} \geq 0$ then*

$$N \leq \mathcal{N}(n, \nu)$$

ii) *If $K_M \geq H$ ($H < 0$) and $q_N \leq d$, then*

$$N \leq \mathcal{N}(n, \nu, Hd^2) .$$

Proof: Take minimal geodesics, γ_i from p to q_i. View $\{\gamma_i'(0)\}$ as a subset of $S^{n-1} \subset M_p^n$. Then Lemma 2.6 gives a lower bound on the distance, θ, between any pair $\gamma_i'(0), \gamma_j'(0)$. The balls of radius $\theta/2$ about the $\gamma_i'(0) \in S^{n-1}$ are mutually disjoint. Hence, if we denote by $V_{n-1,1}(r)$, the volume of a ball of radius r on S^{n-1}, we can take

$$\mathcal{N} = \frac{V_{n-1,1}(\pi)}{V_{n-1,1}(\theta/2)} ,$$

where $V_{n-1,1}(\pi) = \mathrm{Vol}(S^{n-1})$ and θ is the minimum value allowed by Lemma 2.6.

Remark 2.8. It turns out that Corollary 2.7 is the only place in which the hypothesis on *sectional* curvature is used in deriving Gromov's bound on Betti numbers in terms of curvature and diameter. For details, see Theorem 3.8 and §§ 5-6.

The following result is a weak version (with a much shorter proof) of the main result of [CG12], compare also § 8.

Corollary 2.9. *Let M^n be complete,* with $K_{M^n} \geq 0$. *Given p, there exists a compact set C, such that p has no critical points lying outside C. In particular M^n is homeomorphic to the interior of a compact manifold with boundary.*

Proof: The first statement, which is obvious from Corollary 2.7, easily implies the second.

Corollary 2.10. *Let $\mathcal{N}(n, \nu, Hd^2)$ be as in* Corollary 2.7, *and let $r_1 \nu^{\mathcal{N}} < r_2$. Then there exists $(s_1, s_2) \subset [r_1, r_2]$ such that $\rho_p^{-1}((s_1, s_2))$ is free of critical points and*

$$s_2 - s_1 \geq (r_2 - r_1 \nu^{\mathcal{N}})(1 + \nu + \cdots \nu^{\mathcal{N}})^{-1}$$

Moreover, the set of critical points has measure at most $(1 - \nu^{-\mathcal{N}}) r_2$.

Proof: Let $r_1 + \ell_1$ denote the first critical value $\geq r_1$; $\ell_2 + \nu(r_1 + \ell_1)$ the first after $\nu(r_1 + \ell_1)$ etc. It is easy to see that in the worst case

$$\ell_1 = \ell_2 = \cdots = \ell ,$$

$$(\cdots(\nu(\nu(r_1 + \ell) + \ell) + \ell \cdots) + \ell = r_2$$

The first assertion follows easily. The proof of the second is similar.

Remark 2.11. The proof of Corollary 2.7 easily yields an explicit estimate for the constant $\mathcal{N}$. For example, in case $K_{M^n} \geq 0$, we get

$$\mathcal{N}(n,\nu) \leq \left(\frac{\pi}{\frac{1}{2}\cos^{-1}(1/\nu)}\right)^{n-1}$$

Thus, for ν close to 1,

$$\mathcal{N}(n,\nu) \leq \left[\frac{2\pi^2}{(\nu-1)}\right]^{(n-1)/2}$$

3. Background on Finiteness Theorems.

The theorems in question bound topology in terms of bounds on geometry. In subsequent lectures we will prove two such results due to Gromov, [G] and Grove-Petersen [GrP]. Before stating these, we establish the context by giving an earlier result of Cheeger [C1], [C3] (see also [GLP], [GreWu], [Pe1], [Pe2], [We] for related developments).

Theorem 3.1. (Cheeger). *Given $n, d, V, K > 0$, the collection of compact n-manifolds which admit metrics whose diameter, volume and curvature satisfy,*

$$\mathrm{dia}(M^n) \leq d\,,$$
$$\mathrm{Vol}(M^n) \geq V\,,$$
$$|K_M| \leq K\,,$$

contains only a finite number, $C(n,,V^{-1}d^n, Kd^2)$, of diffeomorphism types.

Remark 3.2. The basic point in the proof is to establish a lower bound on the length of a smooth closed geodesic (here one need only assume $K_M \geq K$). This, together with the assumption $K_M \leq K$, gives a lower bound on the injectivity radius of the exponential map (see [CE], Chapter 5). Although Theorem 3.1 predated the use of critical points, the crucial ingredient in the Grove-Petersen theorem below is essentially a generalization of the above mentioned lemma on closed geodesics (compare Example 1.6).

Theorem 3.3 (Grove-Petersen). *Given $n, d, V > 0$ and H, the collection of compact n-dimensional manifolds which admit metrics satisfying*

$$\mathrm{dia}(M^n) \leq d\,,$$
$$\mathrm{Vol}(M^n) \geq V\,,$$
$$K_M \geq H\,,$$

contains only a finite number, $C(n, V^{-1}d^n, Hd^2)$, of homotopy types.

Remark 3.4. In [GrPW], the conclusion of Theorem 3.3 is strengthened to finiteness up to homeomorphism ($n \neq 3$) and up to diffeomorphism ($n \neq 3,4$). The proof employs techniques from "controlled topology". Thus, Theorem 3.3 supersedes Theorem 3.1 (as stated) if $n \neq 3,4$. However, Theorem 3.1 can actually be strengthened to give a conclusion which does not hold under the hypotheses of Theorem 3.3.

Given $\{M_i^n\}$ as in Theorem 3.1, there is a subsequence $\{M_j^n\}$, a manifold M^n, and diffeomorphisms, $\phi_j : M^n \to M_j^n$, such that the pulled back metrics, $\phi_j^*(g_j)$, converge in the $C^{1,\alpha}$-topology, for all $\alpha > 1$ (see the references given at the beginning of this section for further details).

Example 3.5. By rounding off the tip of a cone, a surface of nonnegative curvature is obtained. From this example, one sees that under the conditions of Theorem 3.3, arbitrarily small metric balls need not be contractible. Thus, the criticality radius can be arbitrarily small (compare Remark 1.9).

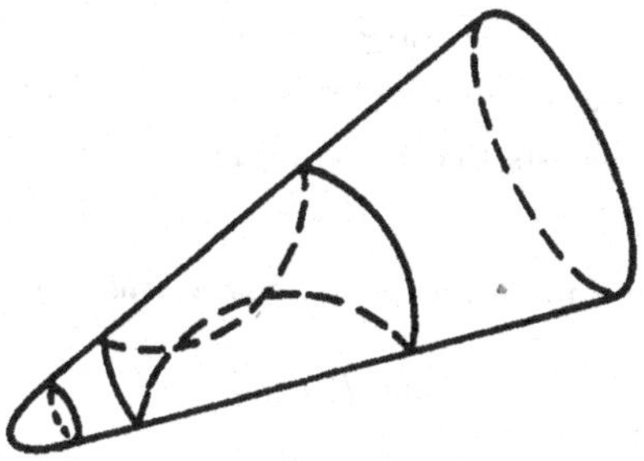

Fig. 3.1

However, it will be shown that the inclusion of a sufficiently small ball into a somewhat larger one is homotopically trivial.

Example 3.6. Consider the *surface* of a solid cylindrical block from which a large number, j, of cylinders (with radii tending to 0) have been removed.

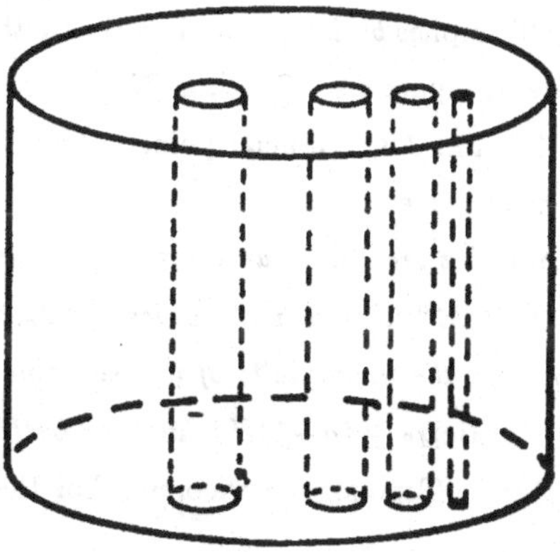

Fig. 3.2

The edges can be rounded so as to obtain a manifold, M_j^2, with $\mathrm{Vol}(M_j^2) \geq V$, $\mathrm{dia}(M_j^2) \leq d$ (but

$\inf K_{M_j^2} \to -\infty$, as $j \to \infty$). For the first Betti number, one has $b^1(M_j^2) = 2j \to \infty$.

Note that the metrics in this sequence can be rescaled so that $K_{M_j^2} \geq -1$, $\mathrm{Vol}(M_j^2) \to \infty$. Then, of course, $\mathrm{dia}(M_j^2) \to \infty$ as well.

Example 3.7. Consider the lens space L_n^3, obtained by dividing

$$S^3 = \{(z_1, z_2) \mid |z_1|^2 + |z_2|^2 = 1\} ,$$

by the action of $\mathbb{Z}_n = \{1, a, \ldots, a^{n-1}\}$, where $a : (z_1, z_2) \to (e^{2\pi i/n} z_1, e^{2\pi i/n} z_2)$. Then $\mathrm{dia}(L_n^3) = 1$, $K_{L_n^3} \equiv 1$, but $\mathrm{Vol}(M_n^3) \to 0$, and $H_1(L_n^3, \mathbb{Z}) = \mathbb{Z}_n$. Thus, if the lower bound on volume is relaxed, there are infinitely many possibilities for the first homology group, H_1. Nonetheless, the following theorem of Gromov asserts that for any *fixed* coefficient field, F, the Betti numbers, $b^i(M^n)$ are bounded independent of F.

Theorem 3.8 (Gromov). *Given $n, d > 0$, H, and a field F,* if

$$\mathrm{dia}(M^n) \leq d ,$$
$$K_{M^n} \geq H ,$$

then

$$\sum_i b^i(M^n) \leq C(n, Hd^2) .$$

Corollary 3.9. *If M^n has nonnegative sectional curvature, $K_{M^n} \geq 0$, then*

$$\sum_i b^i(M^n) \leq C(n) .$$

Remark 3.10. The most optimistic conjecture is that $K_{M^n} \geq 0$ implies $b^i(M^n) \leq \binom{n}{i}$, and hence, $\sum_i b^i(M^n) \leq 2^n$. Note $b^i(T^n) = \binom{n}{i}$ where T^n is a flat n-torus. At present, one knows only that $K_{M^n} \geq 0$ (in fact $\mathrm{Ric}_{M^n} \geq 0$) implies $b^1(M^n) \leq n$. But the method of proof of Theorem 3.8 does not give this sharp estimate; compare also [GLP], p. 72.

In proving Theorems 3.1, 3.3 and 3.8, a crucial point is to bound the number of balls of radius ϵ needed to cover a ball of radius r.

Proposition 3.11 (Gromov). *Let the Ricci curvature of M^n satisfy $\mathrm{Ric}_{M^n} \geq (n-1)H$. Then given $r, \epsilon > 0$ and $p \in M^n$, there exists a covering, $B_r(p) \subset \cup_1^N B_\epsilon(p_i)$, ($p_i \in B_r(p)$) with $N \leq N_1(n, Hr^2, r/\epsilon)$. Moreover, the multiplicity of this covering is at most $N_2(n, Hr^2)$.*

Remark 3.12. The condition $\mathrm{Ric}_{M^n} \geq (n-1)H$ is implied by $K_{M^n} \geq H$, in which case, the bound on N_1 could be obtained from Toponogov's theorem. For the proof of Proposition 3.11, see § 7.

Remark 3.13. The conclusion of Theorem 3.8 (and hence of Corollary 2.7) fails if the hypothesis $K_m \geq H$ is weakened to the lower bound on Ricci curvature, $Ric_{M^n} \geq (n-1)H$; see [An], [ShY].

Remark 3.14. S. Zhu has shown that homotopy finiteness continues to hold for $n = 3$, if the lower bound on sectional curvature is replaced by a lower bound on Ricci curvature; [Z]. Whether or not this remains true in higher dimensions is an open problem.

4. Homotopy finiteness.

Pairs of mutually critical points.

The main point in proving the theorem on homotopy finiteness is to establish a lower bound on the distance between a pair of *mutually critical* points (compare Example 1.6). For technical reasons we actually need a quantitative refinement of the notion of criticality.

Definition 4.1. q is *ϵ-almost critical* with respect to p, if for all $v \in M_q$, there exists γ, minimal from q to p, with $\measuredangle(v, \gamma'(0)) \leq \frac{\pi}{2} + \epsilon$.

Theorem 4.2. *There exist* $\epsilon = \epsilon(n, V^{-1}d^n, Hd^2)$, $\delta = \delta(n, V^{-1}d^n, Hd^2) > 0$, *such that if* $p, q \in M^n$

$$\mathrm{dia}(M^n) \leq d \,,$$

$$\mathrm{Vol}(M^n) \geq V \,,$$

$$K_{M^n} \geq H \,,$$

$$\overline{p,q} < \delta d \,,$$

then at least one of p, q is not ϵ-almost critical with respect to the other.

The proof of Theorem 4.2 uses two results on volume comparison. The first of these, Lemma 4.3, is stated below and proved in the Appendix to this section. The second result, Proposition 4.7 is stated and proved in the Appendix.

For $X \subset Y$ closed, put

$$T_r(X) = \{q \in Y \mid \overline{q,X} < r\}$$

(the case of interest below is $Y = S^{n-1}$, the unit $(n-1)$-sphere).

Recall that the volume of a ball in M_H^n is given as follows. Put

$$A_{n-1,H}(s) = \begin{cases} (\frac{1}{\sqrt{H}} \sin s\sqrt{H}\, s)^{n-1} & H > 0 \\ s^{n-1} & H = 0 \\ (\frac{1}{\sqrt{-H}} \sinh \sqrt{-H}\, s)^{n-1} & H < 0 \end{cases}$$

$$V_{n,H}(r) = v_{n-1} \int_0^r A_{n-1,H}(s)\, ds \,,$$

where $v_{n-1} = V_{n-1,1}(\pi)$ is the volume of the unit $(n-1)$-sphere. Then in M_H^n,

$$\mathrm{Vol}(B_r(\underline{p})) = V_{n,H}(r) \,.$$

Lemma 4.3. *Let $X \subset S^n$ be closed. Then*

a)
$$\frac{\mathrm{Vol}(T_{r_1}(X))}{\mathrm{Vol}(T_{r_2}(X))} \geq \frac{V_{n,1}(r_1)}{V_{n,1}(r_2)} \,.$$

Thus,

b)
$$\frac{\mathrm{Vol}(T_{r_2}(X)) - \mathrm{Vol}(T_{r_1}(X))}{\mathrm{Vol}(T_{r_2}(X))} \le \frac{V_{n,1}(r_2) - V_{n,1}(r_1)}{V_{n,1}(r_2)}\,.$$

Remark 4.4. The lemma actually holds for $X \subset M^n$, where $\mathrm{Ric}_{M^n} \ge (n-1)H$ provided $V_{n,1}(r)$ is replaced by $V_{n,H}$ (see Proposition 7.1).

Proof of Theorem 4.2: By scaling, we can assume $d = 1$.

In i)-iii) below we determine ϵ, δ. In iv) we show that they have the desired properties.

i) Fix $\epsilon > 0$. Let $\underline{\gamma}_0, \underline{\gamma}_1$ determine a hinge in M_H^2 with angle,

$$\alpha < \frac{\pi}{2} - \epsilon\,,$$

at the point,

$$\underline{\gamma}_0(\ell_0) = \underline{\gamma}_1(0)$$

Here, $L[\underline{\gamma}_i] = \ell_i$. Let $\delta = \delta(H, \epsilon, r)$ be the length of the base of an isosceles triangle in M_H^2 with equal sides of length r and angle $\pi/2 - \epsilon$ opposite these sides. Then if

$$\ell_0 \le \delta\,,$$

$$\ell_1 \ge r\,,$$

we have

$$\overline{\underline{\gamma}_0(0), \underline{\gamma}_1(\ell_1)} < \ell_1\,.$$

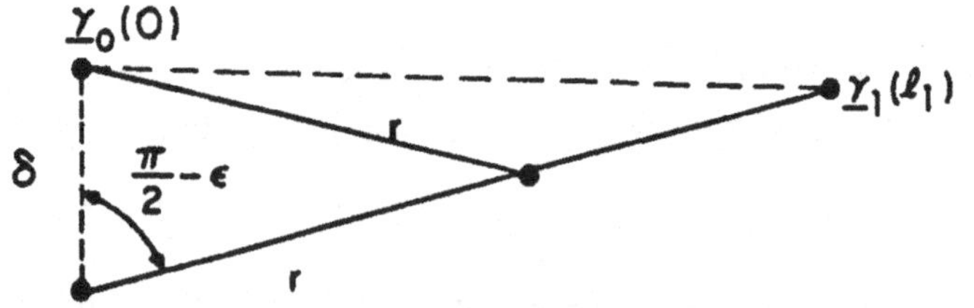

Fig. 4.1

ii) Determine $\epsilon = \epsilon(n, V^{-1}, H)$ by

$$\frac{V_{n-1,1}(\frac{\pi}{2}+\epsilon) - V_{n-1,1}(\frac{\pi}{2}-\epsilon)}{V_{n-1,1}(\frac{\pi}{2}+\epsilon)} \times V_{n,H}(1) = \frac{V}{6}\,.$$

iii) Determine $r = r(n, V^{-1}, H)$ by

$$V_{n,H}(r) = \frac{V}{6}\,.$$

iv) Assume p, q are mutually ϵ-almost critical and that $\overline{p,q} < \delta$ with δ, ϵ, r as in i)-iii). We claim $\mathrm{Vol}(M^n) \leq \frac{2}{3}V$ which is a contradiction.

Let

$$M^n(p) = \{x \in M^n \mid \overline{x,p} < \overline{x,q}\}$$

Since $M \setminus (M^n(p) \cup M^n(q))$ has measure zero, by symmetry, it suffices to show

$$\mathrm{Vol}(M^n(p)) \leq \frac{V}{3}\,.$$

Let $X \subset S^{n-1} \subset M_p^n$ be the set of tangent vectors to minimal geodesics from p to q. By assumption, $\overline{T_{(\pi/2)+\epsilon}(X)} = S^{n-1}$. Hence by ii), Lemma 4.3 b) and Proposition 4.7, the volume of the set of points, $x \in M^n \setminus C_p$, such that $x = \gamma(\ell)$, and $\gamma'(0) \notin T_{(\pi/2)-\epsilon}(X)$, is at most $V/6$.

But if $y = \sigma(u)$, $\sigma'(0) \in T_{(\pi/2)-\epsilon}(X)$ and $u > r$, then by i) and Toponogov's theorem B), we have

$$\overline{q,y} < \overline{p,y}\,.$$

Therefore $y \notin M^n(p)$.

By the choice of r (see iii)) the set of such $y \in M^n(p)$ has volume $\leq V/6$ (see again Proposition 4.7). Thus, we get the contradiction

$$\mathrm{Vol}(M^n(p)) \leq \frac{V}{6} + \frac{V}{6} = \frac{V}{3}\,.$$

Let $\Delta \subset M \times M$ denote the diagonal.

Corollary 4.5. *Let M^n, δ be as above. Then there exists a deformation retraction H_t : $T_{(\delta/2)d}(\Delta) \to \Delta$ $(t \in [0,1])$ such that the curves, $t \to H_t(p,q)$ have length*

$$L[H_t(p,q)] \leq R(n, V^{-1}d^n, Hd^2)\,\overline{p,q}\,.$$

Proof: By scaling, it suffices to assume $d = 1$. Let $(p,q) \in T_{\delta/2}(\Delta)$ with say q not ϵ-critical with respect to p. Let U_q, W_q be as in the Isotopy Lemma 1.4. Let $W'_{(p,q)}$ be the vector field $(0, W_q)$ on some sufficiently small neighborhood $V_p \times W_q$. (By averaging under the flip, we can even replace $W'_{(p,q)}$ by $W''_{(p,q)}$ such that $W''_{(p,q)} = W''_{(q,p)}$). The proof is completed by a partition of unity construction and first variation argument like those in the Isotopy Lemma 1.4 (the deformation we obtain does not necessarily preserve $T_{\delta/2}(\Delta)$, but satisfies the estimate above).

Curves varying continuously with their endpoints.

Let $(p,q) \in T_{\delta/2}(\Delta)$. Write $H_t(p,q) = (\phi_1(t,p,q),\ \phi_2(t,p,q))$. Put

$$\phi(t,p,q) = \begin{cases} \phi_1(2t,p,q) & 0 \leq t \leq \frac{1}{2} \\ \phi_2(1-2t,p,q) & \frac{1}{2} \leq t \leq 1 \end{cases}$$

Then $\phi(t,p,q)$ is a curve from p to q, which depends continuously on (p,q), with

$$L[\phi(t,p,q)] \leq R\overline{p,q} \, .$$

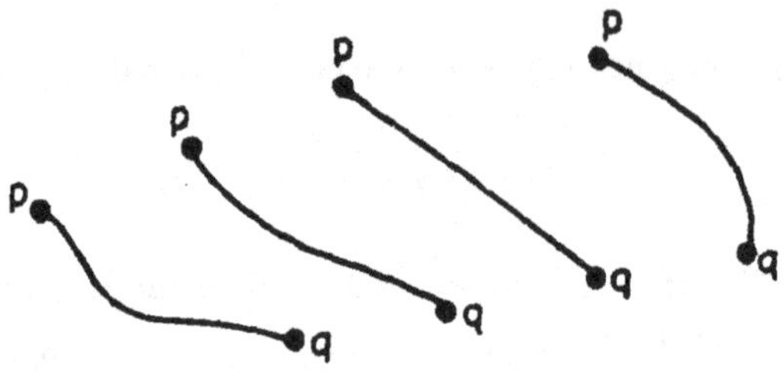

Fig. 4.2

Maps which are close are homotopic.

Corollary 4.6. *Let N be arbitrary and $f_i : N \to M^n$ (as above) $i = 1,2$. If $\overline{f_1(x), f_2(x)} < \frac{\delta}{2}d$, for all $x \in N$, then f_0, f_1 are homotopic.*

Proof: The required homotopy is given by

$$f_t(x) = \phi(t, f_0(x), f_1(x)) \, .$$

Mapping in simplices (center of mass).

Let $(p_0, \ldots, p_k) \in M \times \cdots \times M$, such that $\overline{p_i, p_j} \leq (1 + \cdots + R^{k-1})^{-1}\frac{\delta}{2}d$. Construct a map of a k-simplex into M^n, inductively as follows.

i) Join p_0 to p_1 by $\phi(t, p_0, p_1)$

ii) Join p_2 to each point of $\phi(t, p_0, p_1)$ by $\phi(s, \phi(t, p_0, p_1), p_2)$.

iii) Join p_3 to each point of $\phi(s, \phi(t, p_0, p_1), p_2)$ by $\phi(u, \phi(s, \phi(t, p_0, p_1), p_2), p_3)$, etc.

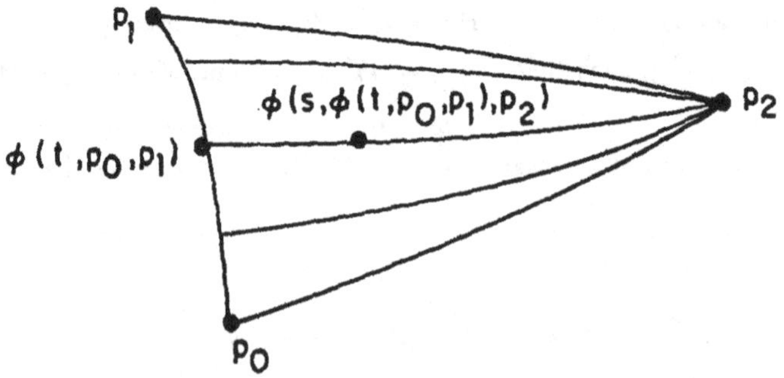

Fig. 4.3

After an obvious reparametrization, we get a map to M^n of the simplex $(\alpha_0, \ldots, \alpha_k)$, $0 \leq \alpha_i \leq 1$, $\sum \alpha_i = 1$.

Proof of Theorem 3.3:

i) By scaling, the metric we can assume $d = 1$

ii) By Proposition 3.11, for any such M^n, we can fix a covering,

$$M^n = \cup B_\epsilon(p_i) \ , \ \text{where } 0 \leq i \leq N_1(n, H, \epsilon, 10) \ .$$

Moreover, the multiplicity of this covering is $\leq N_2(n, H)$. Take

$$\epsilon = \frac{\delta}{12}(1 + R + \cdots + R^{N_1 - 1})^{-1} \ ,$$

with δ, R as in Theorem 4.1.

iii) Since $0 < \overline{p_i, p_j} \leq 1$, by the "pigeon-hole" principle, we can divide the collection of all such M^n into $C(n, V^{-1}, H)$ classes, such that if M_1^n, M_2^n are in the same class,

$$\text{card}(\{p_{i,1}\}) = \text{card}(\{p_{i,2}\}) = c \leq N_1 \ ,$$

and for $\{p_{i,\ell}\} \subset M_\ell^n$, $\ell = 1, 2$ as above,

$$|\overline{p_{i,1}, p_{j,1}} - \overline{p_{i,2}, p_{j,2}}| \leq \frac{\delta}{12}(1 + R + \cdots + R^{N_1 - 1})^{-1} \ .$$

iv) It suffices to show that M_1^n, M_2^n as in iii) are homotopy equivalent. Construct a map $h_1 : M_1^n \to M_2^n$ as follows. Choose a partition of unity, $\sum \phi_i \equiv 1$, subordinate to $\{B_\epsilon(p_{i,1})\}$. Define a map η_1 of M_1^n to the standard $(c-1)$-simplex by

$$\eta_1(x) = (\phi_0(x), \ldots, \phi_{c-1}(x)) \ .$$

Let K be the subcomplex consisting of those closed simplices whose interior has nonempty intersection with range η_1. It follows from iii) that for any $\sigma \in K$ we can define a map $g_1 : \sigma \to M_2^n$ (using the center of mass construction). The maps on the various $\sigma \subset K$ fit together to give $g_1 : K \to M_2^n$. Put $h_1 = g_1 \eta_1$ and define $h_2 : M_2^n \to M_1^n$ similarly.

Let Id_{M_j} denote the identity map on M_j. It is easy to see that the pairs $(h_2 h_1, Id_{M_1})$, $(h_1 h_2, Id_{M_2})$ satisfy the hypothesis of Corollary 4.6. q.e.d.

Appendix

Proof of Lemma 4.3: For all $\epsilon > 0$, we can find a finite set of points, X_ϵ, such that $T_\epsilon(X_\epsilon) \supset X$, $T_\epsilon(X) \supset X_\epsilon$. Since ultimately, we can let $\epsilon \to 0$, it follows easily that it suffices to assume that $X = p_1 \cup \cdots \cup p_N$ is itself a finite set. Fix i and define the *starlike* set, U_r, by

$$U_r = \{x \in T_r(X) \mid \overline{x, p_i} < \overline{x, p_j} \quad \forall j \neq i\} \ .$$

Since i is arbitrary, it clearly suffices to show

$$\frac{\mathrm{Vol}(U_{r_1})}{\mathrm{Vol}(U_{r_2})} \geq \frac{V_{n,1}(r_1)}{V_{n,1}(r_2)} .$$

Put

$$\{A_{r_1} = \gamma(s) \mid \gamma(s) \in U_{r_1} \text{ and } \exists t > r_1 \text{ with } \gamma(t) \in U_{r_2}\} .$$

Then, clearly we get

$$\frac{Vol(U_{r_1})}{Vol(U_{r_2})} \geq \frac{Vol(A_{r_1})}{Vol(U_{r_2})} \geq \frac{V_{n,1}(r_1)}{V_{n,1}(r_2)} . \quad q.e.d.$$

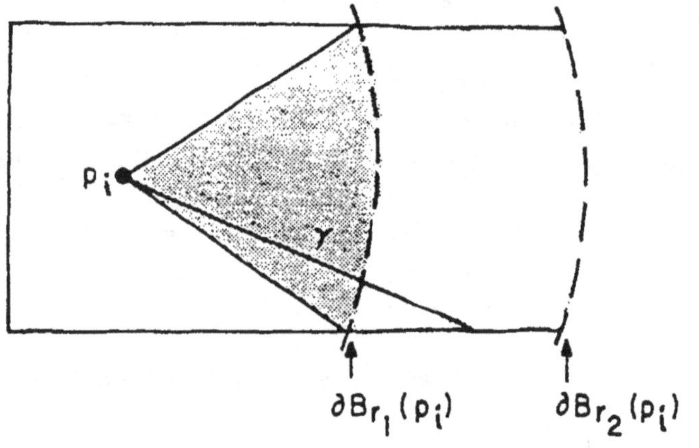

Fig. 4.4

Let γ be a geodesic with $\gamma(0) = p$ and let $\gamma(\ell_j)$ denote the cut point of γ. Let U be the interior of the cut locus in M_p^n. Thus,

$$U = \{t\gamma'(0) \mid 0 \leq t < \ell_j\} .$$

Then $\exp_p \overline{U} = M^n$.

Let $I : M_p^n \to (\underline{M}_H^n)_{\underline{p}}$ be an isometry and put

$$\underline{U} = \exp_{\underline{p}} \cdot I(U) .$$

Proposition 4.7. *The map*

$$\exp_p \cdot I^{-1} \cdot \exp_{\underline{p}}^{-1} \mid \overline{\underline{U}}$$

is distance decreasing if $K_{M^n} \geq H$ and volume decreasing if $\mathrm{Ric}_{M^n} \geq (n-1)H$.

Proof: The first assertion is clear from Toponogov's theorem B). In particular the above map is volume decreasing in this case. For the second assertion, see Remark 7.3.

5. Betti numbers and rank.

Gromov's inequality (Theorem 3.8)

$$\sum b^i(M^n) \leq C(n, Hd^2)$$

depends on a novel method of estimating Betti numbers (as well as on the interaction between curvature and critical point theory, in particular, Corollary 2.7). In the present section, we estimate Betti numbers in terms of an invariant called *rank*. This part of the discussion (and much of that of §6) applies to metric spaces considerably more general than riemannian manifolds. In §6, we show that in the context of Theorem 3.8, rank can be estimated in terms of curvature and diameter (specifically in terms of the numbers $\mathcal{N}(n, Hd^2)$ of Corollary 2.7 and $N_1(n, Hd^2, 10^{n+1})$ of Proposition 3.11).

Let $U_1, U_2 \subset M$ be open. The Mayer-Vietoris sequence,

$$\to H^{i-1}(U_1 \cap U_2) \to H^i(U_1 \cup U_2) \to H^i(U_1) \oplus H^i(U_2) \to$$

leads immediately to the estimate

$$b^i(U_1 \cup U_2) \leq b^i(U_1) + b^i(U_2) + b^{i-1}(U_1 \cap U_2)$$

(we regard $b^j(X) = 0$ for $j < 0$).

This generalizes as follows:

Consider $U_1, \ldots, U_N$ and put

$$U_{(j)} := U_{k_0} \cap \cdots \cap U_{k_j}$$

Proposition 5.1.

$$b^i(U_1 \cup \cdots \cup U_N) \leq \sum_{U_{(j)}} b^j(U_{(i-j)})$$

Proof: Note that

$$U_0 \cup \ldots \cup U_{t+1} = (U_1 \cup \cdots U_t) \cup U_{t+1}$$

$$(U_0 \cup \cdots \cup U_t) \cap U_{t+1} = (U_0 \cap U_{t+1}) \cup \ldots \cup (U_t \cap U_{t+1})$$

Apply the previous estimate and induction to the pair $(U_1 \cup \ldots \cup U_t)$, U_{t+1} and use induction to estimate $b^{i-1}((U_0 \cup \ldots \cup U_t) \cap U_{t+1})$.

It is extremely useful to further generalize these estimates to give bounds on the ranks of induced maps on cohomology (note $b^i = \mathrm{rk}(\mathrm{Id}_{H^i})$).

Let $V_1 \xrightarrow{g} V_2 \xrightarrow{f} V_3$ be linear transformations of vector spaces. Then

$$rk(fg) \leq \min(rk(f), rk(g)) .$$

Thus, if $A\overset{u}{\subset}B\overset{v}{\subset}C\overset{w}{\subset}D$, with u, v, w, the inclusion maps and u^*, v^*, w^*, the induced maps on cohomology, then

$$(*)\qquad rk((uvw)^*) \le rk(v^*)\ ,$$

Definition 5.2. If $A \subset B$ let $b^i(A, B)$ denote $rk(u^*)$, where $u^* : H^i(B) \to H^i(A)$.

Remark 5.3. If A, B are open, with $\overline{A} \subset B$, then there exists a submanifold, Y^n, with smooth boundary, such that $A \subset Y^n \subset B$. Then

$$b^i(A, B) \le b^i(Y, Y) = b^i(Y) < \infty\ .$$

Let $\overline{U}_i^j \subset U_i^{j+1}$, $i = 1, \dots, N$, $j = 0, \dots, n+1$. Put $X^j = \cup_i U_i^j$. Thus

$$X^0 \subset X^1 \subset \dots \subset X^{n+1}\ .$$

Then we have the following generalization of Proposition 5.1.

Proposition 5.4.

$$b^i(X^0, X^{n+1}) \le \sum_{j,(i-j)} b^j(U^j_{(i-j)}, U^{j+1}_{(i-j)})$$

The proof of Proposition 5.4, which is a standard application of the *double complex* associated to an open cover, is given in the Appendix to this section.

We are particularly interested in coverings by balls. First we note the obvious

Lemma 5.5.

$$B_r(p_1) \cap \dots \cap B_r(p_j) \ne \emptyset$$

implies that for $1 \le i \le j$

$$B_r(p_1) \cap \dots \cap B_r(p_j) \subset B_r(p_i) \subset B_{5r}(p_i) \subset B_{10r}(p_1) \cap \dots \cap B_{10r}(p_j)\ .$$

Proof: This follows immediately from the triangle inequality.

Content.

Put

$$b^i(r, p) := b^i(B_r(p),\ B_{5r}(p))$$

$$\text{cont}(r, p) := \sum_i b^i(r, p)$$

Note. If $r > \text{dia}(M)$, then $b^i(r, p) = b^i(M)$.

Corollary 5.6.

$$b^i(B_r(p_1) \cap \cdots B_r(p_j)\ ,\ B_{10\,r}(p_1) \cap \cdots B_{10\,r}(p_j)) \le b^i(r, p_i)\ , \quad 1 \le i \le j\ .$$

Proof: This follows from Lemma 5.5 and the inequality $(*)$ preceding Definition 5.2.

Assume now that for any ball, $B_r(p)$, we have $B_r(p) \subset \cup_1^N B_\epsilon(p_i)$, with $p_i \in B_r(p)$ and $N \le N(\epsilon, r)$.

For the next corollary we will need the observation that if $p' \in B_r(p)$, then by the triangle inequality,

$$(+) \qquad B_{10 \cdot r/10}(p') = B_r(p') \subset B_{5r}(p) .$$

Corollary 5.7. *If for all $p' \in B_r(p)$ and $j = 1, \dots, n+1$*

$$\mathrm{cont}(10^{-j} r, p') \le c ,$$

then

$$\mathrm{cont}(r, p) \le (n+1) \cdot 2^{N(10^{-(n+1)} r, r)} \cdot c$$

Proof: Take a cover of $B_r(p)$ by balls $B_{10^{-(n+1)} r}(p_i)$ as above. Put $U_i^j = B_{10^{j-(n+1)} r}(p_i)$ and apply Proposition 5.4. The total number of intersections, $U^j_{(i-j)}$, on the right-hand side of the inequality in Proposition 5.4 is at most $(n+1) 2^{N(10^{-(n+1)} r, r)}$. Also, by Corollary 5.6, we certainly have

$$b^j(U^j_{(i-j)}, U^{j+1}_{(i-j)}) \le c .$$

Since

$$B_r(p) \subset U^0 \subset U^{n+1} \subset B_{5r}(p)$$

(see (+) above) the claim follows from the inequality (*).

Thus, the content of a given ball can be estimated in terms of the contents of certain smaller balls.

There is an easier, but equally important means of estimating content.

Compression

Definition 5.8. We say $B_r(p)$ compresses to $B_s(q)$ and write $B_r(p) \mapsto B_s(q)$, if

1) $5s + \overline{p,q} \le 5r$.

2) There is a homotopy, $f_t : B_r(p) \to B_{5r}(p)$, with f_0 the inclusion and $f_1(B_r(p)) \subset B_s(q)$.

Note. By 1), $B_r(p) \mapsto B_s(q)$ implies $s \le r$.

Lemma 5.9. *If $B_r(p) \mapsto B_s(q)$, then*

$$b^i(r, p) \le b^i(s, q) .$$

$$\mathrm{cont}(r, p) \le \mathrm{cont}(s, q) ,$$

Proof: Obvious by (*).

Rank

Now for each ball, $B_r(p)$ we define (inductively) an integer called the rank. This invariant enables us to conveniently combine our two methods of estimating content (Corollary 5.7 and Lemma 5.9).

Definition 5.10.

i) $\text{rank}(r,p) := 0$, if $B_r(p) \mapsto B_s(q)$, with $B_s(q)$ contractible.

ii) $\text{rank}(r,p) := j$, if $\text{rank}(r,p)$ is *not* $\leq j-1$ and if $B_r(p) \mapsto B_s(q)$, such that for all $q' \in B_s(q)$, with $s' \leq \frac{1}{10}s$, we have $\text{rank}(s',q') \leq j-1$.

Remark 5.11. Of necessity, there exists *some* $B_{s'}(q')$ as in ii) with $\text{rank}(s',q') = j-1$; otherwise we would have $\text{rank}(r,p) \leq j-1$.

Proposition 5.12. *If balls of radius* $< \epsilon$ *are contractible, then*

$$\text{rank}(r,p) \leq \frac{\log r/\epsilon}{\log 10} + 1 .$$

Proof: Trivial by induction.

Corollary 5.13.

$$\text{cont}(r,p) \leq \left((n+1)\,2^{N(10^{-(n+1)}r,r)}\right)^{\text{rank}(r,p)} .$$

Proof: By induction, this follows from Corollary 5.7, and Definition 5.10.

Remark 5.14. Needless to say, there is some degree of arbitrariness involved in the choice of constants 5, 10, (and $\frac{1}{2}$) which appear in Definitions 5.2, 5.10 (and 6.1) respectively.

Remark 5.15. In certain respects, our terminology and notation in §§5,6, differ somewhat from that of [G].

Appendix. The generalized Mayer-Vietoris estimate.

Our considerations here are very similar to those of [BT], Chapter II.

Let $\cup_i U_i = X$ be an open cover of X. Put

$$C^{i,j} = \oplus_{(i)} C^j(U_{(i)})$$

where $C^j(U_{(i)})$ denotes the space of singular j-cochains, with coefficients in some field. The *double complex* $C^* := \oplus_{i,j} C^{i,j}$ has two differentials,

$$\delta : C^{i,j} \to C^{i+1,j} , \quad \delta^2 = 0$$
$$d : C^{i,j} \to C^{i,j+1} , \quad d^2 = 0$$

(d is induced by $d : C^j(U_{(i)}) \to C^{j+1}(U_{(i)})$). The *total differential,* $(d+\delta)$, also satisfies $(d+\delta)^2 = 0$.

A j-cochain on $X = \cup U^i$ determines and is determined by $x \in C^{0,j}$, with $\delta x = 0$. Under this identification $d : C^{0,j} \to C^{0,j+1}$ corresponds to $d : C^j(X) \to C^{j+1}(X)$.

A basic fact we need is that C^* is *δ-acyclic for $i > 0$*, i.e. $y \in C^{i,j}$ $(i > 0)$; $\delta y = 0$ implies $y = \delta z$ for some $z \in C^{i-1,j}$.

Let $X^k = \cup_i U_i^k$, where $\overline{U}_i^k \subset U_i^{k+1}$. Denote by $C^*(k)$, the double complex associated to $X^k = \cup_i U_i^k$. There is a natural restriction map $r_k : C^{i,j}(k) \to C^{i,j}(k+1)$, commuting with d, δ.

Proof of Proposition 5.4: Let $Z^j \subset C^{0,j}(j+1) \cap \ker d + \delta$, be a space of representative cocycles, mapping isomorphically onto $H^j(X^{j+1})$. We will define a filtration,

$$Z^j = Z^j_{j+1} \supset Z^j_j \supset \ldots \supset Z^j_0 ,$$

such that

$$(\times) \qquad \dim(Z^j_{s+1} / Z^j_s) \leq \sum_{s,(j-s)} b^s(U^s_{(j-s)}, U^{s+1}_{(j-s)}) ,$$

and if $z \in Z^j_0$, then $r_0^* \cdot r_1^* \cdots r_j^*(z)$ is *exact*. This will suffice to prove Proposition 5.4.

Put

$$Z^j_j := \{z \in Z^j \mid r_j^*(z) \text{ is } d \text{ exact}\} .$$

Choose a linear map $d^{-1}{:}r_j^*(Z^j_j) \to C^{0,j-1}(j)$, $dd^{-1}(z) = z$. From $\delta z = 0$, $d\delta = -\delta d$, we get $d\delta d^{-1}(z) = 0$ (and of course $\delta(\delta d^{-1}(z)) = 0$).

Define $Z^j_{j-1} \subset Z^j_j$ by

$$Z^j_j := \{z \in Z^j_{j-1} \mid r_{j-1}^* \delta d^{-1} r_j^*(z) \text{ is } d \text{ exact } \}$$

By proceeding in this way, we obtain $Z^j_{j+1} \supset Z^j_j \supset \ldots \supset Z^j_0$, for which the inequality $(\times)$ obviously holds.

Note that $z \in Z^j_0$ implies

$$r_0^* \delta d^{-1} r_1^* \delta d^{-1} \ldots \delta d^{-1} r_j^*(z) \equiv 0 ,$$

since an exact 0-cochain vanishes identically.

To show $r_0^* \ldots r_j^*(z)$ is exact, put

$$a_s = (-1)^{s+j-1} r_0^* \cdots r_{s-1}^* (d^{-1} r_s^*)(\delta d^{-1} r_{s+1}^* \cdots \delta d^{-1} r_j^*)(z) .$$

Then

$$r_0^* \ldots r_j^*(z) = (d+\delta)(a_{j-1} + \ldots + a_0) .$$

Using δ-acyclicity, choose $b_0 \in C^{j-1,0}(0)$, with $\delta b_0 = a_0$. Put $a_1' = a_1 - db_0$. Then

$$\begin{aligned} r_0^* \ldots r_j^*(z) &= (d+\delta)(a_{j-1} + \ldots + a_0 - (d+\delta)b_0) \\ &= (d+\delta)(a_{j-1} + \ldots + a_2 + a_1') \end{aligned}$$

Proceeding in this way, we find by induction, $\hat{a}_{j-1} \in C^{0,j-1}(0)$, with

$$r_0^* \cdot \ldots r_j^*(z) = (d+\delta)\hat{a}_{j-1} \, .$$

Then, we have

$$r_0^* \cdots r_j^*(z) = d\hat{a}_{j-1} \, ,$$
$$0 = \delta\hat{a}_{j-1} \, ,$$

which completes the proof.

6. Rank, curvature and diameter.

In this section, we will show how a lower bound on curvature leads to an estimate on $\operatorname{rank}(r,p)$, and hence via Corollary 5.13, to an estimate on $\operatorname{cont}(r,p)$. For this purpose, it is convenient to work with a slightly modified definition of rank.

Definition 6.1. A ball, $B_r(p)$, is called *incompressible* if $B_r(p) \mapsto B_s(q)$ implies $s > \frac{1}{2}r$.

It is obvious that any ball, $B_r(p)$, can be compressed either to a contractible ball (in which case $\operatorname{rank}(r,p) = 0$) or to an incompressible ball.

Definition 6.2.

i) $\operatorname{rank}'(p,r) := 0$, if $B_r(p) \mapsto B_s(q)$, with $B_s(q)$ contractible

ii) $\operatorname{rank}'(p,r) := j$ if $\operatorname{rank}'(p,r) \neq j-1$ and $B_r(p) \mapsto B_s(q)$ such that: $B_s(q)$ is incompressible and for all $q' \in B_s(q)$ and $s' \leq \frac{1}{10}s$, we have $\operatorname{rank}(q',s') \leq j-1$.

Thus, we have modified Definition 5.10, by adding the stipulation that the ball, $B_s(q)$, of ii), must be *incompressible.*

Clearly, $\operatorname{rank}'(p,r)$ still satisfies the bound of Proposition 5.12. Moreover, it is obvious that

$$\operatorname{rank}(r,p) \leq \operatorname{rank}'(r,p) \, .$$

The reason for insisting on incompressibility in the definition of $\operatorname{rank}'(r,p)$ stems from

Lemma 6.3. *Let $B_r(p) \subset M^n$, a complete riemannian manifold. Assume*

$$5s + \overline{p,y} \leq 5r \, ,$$
$$\overline{p,y} \leq 2r \, .$$

Then if $B_r(p)$ does not compress to $B_s(y)$, there exists a critical point, x, of y, with

$$s \leq \overline{x,y} \leq r + \overline{p,y} \, .$$

Thus, $x \subset B_{r+2\overline{p,y}}(p) \subset B_{5r}(p)$.

Proof: If there were no such critical point, then by the Isotopy Lemma 1.4, the ball $B_{r+\overline{p,y}}(y)$ could be deformed to lie inside of $B_s(y)$. Since, $5s + \overline{p,y} \leq 5r$, and

$$B_r(p) \subset B_{r+\overline{p,y}}(y) \subset B_{5r}(p) \, ,$$

this would contradict the assumption that $B_r(p)$ does not compress to $B_s(y)$.

Now we can show a connection between the size of $\mathrm{rank}'(r,p)$ and the existence of critical points.

We will need the observation that if $p' \in B_r(p)$, then by the triangle inequality,

$$(*) \qquad B_{5\cdot(r/10)}(p') = B_{r/2}(p') \subset B_{3r/2}(p) .$$

Lemma 6.4. *Let M^n be riemannian and let* $\mathrm{rank}'(r,p) = j$. *Then there exists $y \in B_{5r}(p)$ and $x_j, \dots, x_1 \in B_{5r}(p)$, such that for all $i \leq j$, x_i is critical with respect to y and*

$$\overline{x_i, y} \geq \frac{5}{4}\overline{x_{i-1}, y} .$$

Proof: We can assume without loss of generality that $B_r(p)$ is incompressible (in this case, we will see that $y \in B_{3r/2}(p)$, $x_i \in B_{3r/2}(p)$). Put $p_j = p$, $r_j = r$. By the definition of $\mathrm{rank}'(r,p)$, there exists $\hat{p}_{j-1} \in B_{r_j}(p_j)$, $\hat{r}_{j-1} \leq \frac{1}{10}r_j$, such that

$$\mathrm{rank}'(\hat{r}_{j-1}, \hat{p}_{j-1}) = j-1 .$$

By $(*)$ above,

$$B_{3r_j/2}(p_j) \supset B_{5r_{j-1}}(\hat{p}_{j-1}) .$$

If $B_{\hat{r}_{j-1}}(\hat{p}_{j-1})$ is incompressible, put

$$p_{j-1} = \hat{p}_{j-1} , \quad r_{j-1} = \hat{r}_{j-1} .$$

If not, there exists an incompressible ball, which in this case we call $B_{r_{j-1}}(p_{j-1})$, such that $B_{\hat{r}_{j-1}}(\hat{p}_{j-1}) \mapsto B_{r_{j-1}}(p_{j-1})$ and

$$\mathrm{rank}'(r_{j-1} , p_{j-1}) = j-1 .$$

Since $B_{\hat{r}_{j-1}}(\hat{p}_{j-1}) \mapsto B_{r_{j-1}}(p_{j-1})$ implies $B_{5r_{j-1}}(p_{j-1}) \subset B_{5\hat{r}_{j-1}}(\hat{p}_{j-1})$, in either case we obtain

$$B_{3r_j/2}(p_j) \supset B_{5r_{j-1}}(p_{j-1})$$

Also, since in the second case $r_{j-1} \leq \hat{r}_{j-1}$, in either case, we have

$$r_{j-1} \leq \frac{1}{10}r_j .$$

By proceeding in this fashion, we obtain balls, $B_{r_i}(p_i)$, $i = 0, 1, \dots, j$, such that for $1 \leq i \leq j$, $B_{r_i}(p_i)$ is incompressible and

$$B_{3r_i/2}(p_i) \supset B_{5r_{i-1}}(p_{i-1})$$
$$r_{i-1} \leq \frac{1}{10}r_i .$$

Put $y = p_0$. Then, $y \in B_{3r_i/2}(p_i)$, for all $1 \le i \le j$. In particular,

$$\overline{p_i, y} + 5 \cdot \frac{1}{2} r_i \le 4r_i < 5r_i \,,$$
$$\overline{p_i, y} \le \frac{3}{2} r_i < 2r_i$$

(the conditions of Lemma 6.3).

Since, $B_{r_i}(p_i)$ is incompressible, it does not compress to $B_{r_i/2}(y)$. Thus, by Lemma 6.3, there exists a critical point, x_i, with

$$\frac{1}{2} r_i \le \overline{x_i, y} \le r_i + 2 \cdot \frac{3}{2} r_i = 4r_i \,.$$

Then,

$$\begin{aligned} \overline{x_i, y} &\ge \frac{1}{2} r_i \\ &\ge 5r_{i-1} \\ &\ge \frac{5}{4} 4r_{i-1} \\ &\ge \frac{5}{4} \overline{x_{i-1}, y} \,. \quad \text{q.e.d.} \end{aligned}$$

Corollary 6.5.

$$\operatorname{rank}(r, p) \le \begin{cases} \mathcal{N}(n) & H = 0 \\ \mathcal{N}(n, Hd^2) & H < 0 \,. \end{cases}$$

Proof: This follows immediately from Corollary 2.7, Lemma 6.4 and the inequality $\operatorname{rank}(r, p) \le \operatorname{rank}'(r, p)$.

Proof of Theorem 3.8: By Proposition 3.11,

$$N(10^{-(n+1)} r, r) \le \begin{cases} N_1(n,\, 10^{-(n+1)}) \\ N_1(n,\, Hd^2,\, 10^{-(n+1)}) \end{cases}$$

with $N_1(10^{-(n+1)} r, r)$ the covering number appearing in Corollary 5.7. Hence, by that corollary, and by Corollary 6.5 for all $\epsilon > 0$,

$$\begin{aligned} \sum b^i(M^n) &= \operatorname{cont}(d + \epsilon, p) \\ &\le ((n+1) 2^{N_1})^{\mathcal{N}} \end{aligned}$$

q.e.d.

Remark 6.6. Inspection of the bounds given in Corollary 2.7 and Proposition 3.11 (compare Corollary 2.11) reveals that the dependence on n of the constant $C(n)$ in Theorem 3.8 is at worst of the form $2^{2^{an}}$ (for suitable $a > 0$). However, Abresch has shown that by arguing more carefully (along essentially the same lines as we have done) one obtains $C(n) \le 2^{an^3}$; [A], [Me]. Recall that in view of the existence of flat tori, $C(n) = 2^n$ is the best one could hope for.

7. Ricci curvature, volume and the Laplacian.

In this section we present some basic properties of manifolds whose Ricci curvature satisfies $\mathrm{Ric}_{M^n} \geq (n-1)H$. In particular, after proving Proposition 3.11, we derive some estimates involving the Laplacian, which are used in §8. There, we prove a theorem of Abresch-Gromoll, asserting that complete manifolds with $\mathrm{Ric}_{M^n} \geq 0$, which satisfy certain additional conditions, have finite topological type.

Let $\{e_i\}$ be an orthonormal basis of M^n_p. We denote by $\mathrm{Ric}(u,v)$ the symmetric bilinear form,

$$\mathrm{Ric}(u,v) = \sum_i \langle R(e_i,u)v, e_i\rangle \,.$$

Thus, $\mathrm{Ric}(u,v)$ is the trace of the linear transformation $w \to R(w,u)v$.

We write

$$\mathrm{Ric}_{M^n} \geq (n-1)H \,,$$

if

$$\mathrm{Ric}(v,v) \geq (n-1)H \,,$$

for all unit tangent vectors v. Of course, this condition is implied by $K_{M^n} \geq H$, but not vice versa.

Suppose $\gamma|[0,\ell]$ contains no cut point. Then the distance function, $r = \rho_{\gamma(0)}$, is smooth near $\gamma\,|\,[0,\ell]$. Put $N = \mathrm{grad}\, r$. Thus, $N(\gamma(t)) = \gamma'(t)$. Let $e_2, \cdots e_n$ be orthonormal, with

$$\langle e_i, \gamma'(0)\rangle_{\gamma(0)} = 0 \,,$$

$$\nabla_N e_i = 0 \,.$$

Then

$$\mathrm{Ric}(N,N) = \sum_i \langle (\nabla_{e_i}\nabla_N - \nabla_N\nabla_{e_i} - \nabla_{[e_i,N]})N, e_i\rangle \,.$$

We have,

$$\nabla_N N \equiv 0 \,.$$

Also,

$$\begin{aligned} -\sum_i \langle \nabla_N \nabla_{e_i} N, e_i\rangle &= -\sum_i N\langle \nabla_{e_i} N, e_i\rangle \\ &= -m' \,, \end{aligned}$$

where m is the mean curvature of the distance sphere, $\partial B_r(\gamma(0))$, in the direction of the inner normal, $-N$ (and $m' = \dfrac{\partial m}{\partial r}$). Finally,

$$\begin{aligned} -\sum \langle \nabla_{[e_i,N]} N, e_i\rangle &= -\sum_{i,j} \langle \nabla_{e_i} N, e_j\rangle \, \langle \nabla_{e_j} N, e_i\rangle \,, \\ &= -\|\mathrm{Hess}_r\|^2 \,, \end{aligned}$$

where Hess_r denotes the Hessian of r. Thus, we get the basic equation,

(*) $$\|\mathrm{Hess}_r\|^2 + \mathrm{Ric}(N,N) = -m' \,.$$

Additionally, letting Δ denote the Laplacian, we have

$$\begin{aligned}\Delta r &= \Big(\sum_{2}^{n} e_i e_i + NN - \nabla_{e_i} e_i - \nabla_N N\Big) r \\ &= m\ .\end{aligned}$$

Alternatively, this relation follows from the formula, for Δ in geodesic polar coordinates,

$$\Delta = \frac{\partial^2}{\partial r^2} + m(x)\frac{\partial}{\partial r} + \widetilde{\Delta}\ ,$$

where $\widetilde{\Delta}$ is the intrinsic Laplacian of the distance sphere $\partial B_r(\gamma(\sigma))$.

An invariant definiton of the Laplacian is,

$$\Delta f = \mathrm{tr}(\mathrm{Hess}_f)\ .$$

Thus, we also have

$$\mathrm{tr}(\mathrm{Hess}_r) = m\ .$$

From the Schwarz inequality and the fact that one eigenvalue of Hess_r is $\equiv 0$ (corresponding to the eigenvector, N), we get

$$(**) \qquad \|\mathrm{Hess}_r\|^2 \geq \frac{m^2}{n-1}\ .$$

Substituting $(*)$ into $(**)$ gives the differential inequality,

$$\mathrm{Ric}(N,N) \leq -\frac{m^2}{n-1} - m'\ .$$

Note that, as $r \to 0$,

$$m(r) \sim \frac{n-1}{r}\ .$$

Set $u = \dfrac{n-1}{m}$. Then if we assume $\mathrm{Ric}_{M^n} \geq (n-1)H$, we easily obtain

$$\frac{u'}{1+Hu^2} \geq 1\ .$$

By integrating this expression, we find that $\mathrm{Ric}_{M^n} \geq (n-1)H$ implies

$$(+) \qquad m(x) \leq m_H(r(x))\ ,$$

or equivalently,

$$(++) \qquad \Delta r(x) \leq \Delta_H r\big|_{r=r(x)}\ ,$$

where $m_H(r)$, the mean curvature of $\partial B_r(\underline{p}) \subset M_H^n$ in direction $-N$, is given by

$$m_H(r) = (n-1)\begin{cases} \sqrt{H}\cot\sqrt{H}\,r & H > 0 \\ r^{-1} & H = 0 \\ \sqrt{-H}\coth\sqrt{-H}\,r & H < 0 \end{cases}$$

Let ω denote the volume form on the unit sphere. Write

$$dr \wedge \mathcal{A}(r)\omega$$

for the volume form on $M^n \setminus C_p$, Then $m = \frac{\mathcal{A}'}{\mathcal{A}}$. Differentiating $\mathcal{A}/\mathcal{A}_{H,n-1}$ and using (+) gives

$$[\mathcal{A}(r)/\mathcal{A}_{H,n-1}(r)] \downarrow ,$$

$$\mathcal{A}(r) \leq \mathcal{A}_{H,n-1}(r) .$$

($\mathcal{A}_{H,n-1}(r)$ is defined prior to Lemma 4.3.)

Now, by arguing as in the proofs, of Lemma 4.3 and Proposition 4.7, we can immediately extend Lemma 4.3 as follows.

Proposition 7.1. *Let* $\mathrm{Ric}_{M^n} \geq (n-1)H$ *and let* $X \subset M^n$ *be compact. Then for* $r_1 < r_2$,

$$\frac{\mathrm{Vol}(T_{r_1}(X))}{\mathrm{Vol}(T_{r_2}(X))} \geq \frac{V_{n,H}(r_1)}{V_{n,H}(r_2)} .$$

Remark 7.2. In the basic case, $X = p$, the above inequality was emphasized in [G]; compare also [C2].

Proof of Proposition 3.11: Take a maximal set of points, p_i, in $\overline{B_{r-\epsilon/2}(p)}$ at mutual distance $\geq \frac{\epsilon}{2}$. Clearly $\{p_i\}$ is $\frac{\epsilon}{2}$-dense in $\overline{B_{r-\epsilon/2}(p)}$, and hence, ϵ-dense in $B_r(p)$. The balls $\{B_{\epsilon/4}(p_i)\}$ are all disjoint. Moreover, by Proposition 7.1,

$$\frac{V_{n,H}(\epsilon/4)}{V_{n,H}(2r)} \leq \frac{\mathrm{Vol}(B_{\epsilon/4}(p_i))}{\mathrm{Vol}(B_{2r}(p_i))} ,$$

while since $B_r(p) \subset B_{2r}(p_i)$,

$$\frac{\mathrm{Vol}(B_{\epsilon/4}(p_i))}{\mathrm{Vol}(B_{2r}(p_i))} \leq \frac{\mathrm{Vol}(B_{\epsilon/4}(p_i))}{\mathrm{Vol}(B_r(p))} .$$

Thus, the number of balls is bounded by

$$\frac{V_{n,H}(2r)}{V_{n,H}(\epsilon/4)} .$$

If $B_\epsilon(p_j) \cap B_\epsilon(p_i) \neq \emptyset$, then $B_\epsilon(p_j) \subset B_{3\epsilon}(p_i)$. Then, as above, it follows that the multiplicity of our covering is bounded by

$$\frac{V_{n,H}(3\epsilon)}{V_{n,H}(\epsilon/2)} . \qquad \text{q.e.d.}$$

Now, the scale invariant inequalities of Proposition 3.11 follow by an obvious scaling argument.

Remark 7.3. At this point it is clear that the hypothesis, $\mathrm{Ric}_{M^n} \geq (n-1)H$, implies that the map in Proposition 4.7 is volume decreasing.

We now observe that the inequality, (++), on the Laplacian of the distance function, can be generalized in a meaningful way so as to include points which lie on the cut locus. This discussion goes back to a fundamental paper of E. Calabi, [Ca].

First we need some definitions.

Definition 7.4. An *upper barrier* for a continuous function f at the point x_0, is a C^2 function, g, defined in some neighborhood of x_0, such that $g \geq f$ and $g(x_0) = f(x_0)$.

The crucial observation for applications to geometry is the following.

Lemma 7.5. *If $\gamma(\ell)$ is a cut point of $\gamma(0)$, then for all $\epsilon < \ell$, $\rho_{\gamma(\epsilon)}(x) + \epsilon$ is an upper barrier for $r = \rho_{\gamma(0)}$ at $\gamma(\ell)$.*

Proof: This follows immediately from the triangle inequality.

Definition 7.6. We say $\Delta f(x_0) \leq a$ (and $\Delta(-f(x_0)) \geq -a$) in the *barrier sense* if for all $\epsilon > 0$, there is an upper barrier $f_{x_0,\epsilon}$ for f at x_0 with

$$\Delta f_{x_0,\epsilon} \leq a + \epsilon \,.$$

Now we can generalize (++) above as follows.

Proposition 7.7. *Let M^n be complete, with*

$$\mathrm{Ric}_{M^n} \geq (n-1)H \,.$$

i) *If $f(r)$ satisfies, $f' \geq 0$, then in the barrier sense,*

$$\Delta f(r(x)) \leq \Delta_H f(r)\big|_{r=r(x)} \,.$$

ii) *If $f(r)$ satisfies $f' \leq 0$, then in the barrier sense,*

$$\Delta f(r(x)) \geq \Delta_H f(r)\big|_{r=r(x)} \,.$$

Proof: It suffices to prove i). At smooth (i.e. noncut) points, it is clear from (+) and the formula for Δ in polar coordinates. At cut points, it follows immediately by using the barrier $f(\rho_{\gamma(\epsilon)}(x) + \epsilon)$.

Functions which satisfy say $\Delta f \geq 0$ in the barrier sense, also satisfy a maximum principle. This fact (due to Calabi) was used by Eschenberg and Heintze [EH] to give a very short proof of the splitting theorem of [CGl1]. (They also gave a somewhat longer, but completely elementary proof along closely related lines). Theorem 7.9 below, which is crucial for the discussion of §8, was partly inspired by their work.

Theorem 7.8 (Maximum principle). *Let M be a connected riemannian manifold and let $f \in C^0(M)$. Suppose that $\Delta f \geq 0$ in the barrier sense. Then f attains no weak local maximum unless it is a constant function.*

For completeness, in the Appendix to this section, we give a proof of Theorem 7.8.

We now give an estimate of Abresch-Gromoll on the growth of nonnegative Lipschitz functions whose Laplacian is bounded above in the barrier sense. This will be applied to excess functions, i.e. functions of the form

$$e(x) = \overline{x,p} + \overline{x,q} - \overline{p,q}$$

(for fixed p, q). The estimate involves a comparison function on the model space M^n_H. We now specify this function.

Given $\underline{y} \in M^n_H$, as usual, put $r(\underline{x}) := \rho_y(\underline{x})$. Fix $R > 0$ and a constant $b > 0$. Then there is a unique smallest function, $G(r(\underline{x}))$, defined on $M^n_H \setminus \underline{y}$ satisying,

1) $G > 0$ $(0 < r < R)$

2) $G' < 0$ $(0 < r < R)$

3) $G(R) = 0$,

4) $\Delta_H G \equiv b$,

For any H, the function G can be written in closed form; see [AGl]. Here we need only the case, $H = 0$, $n > 2$. Then,

$$G(r) = \frac{b}{2n}(r^2 + \frac{2}{n-2}R^n r^{2-n} - \frac{n}{n-2}R^2)\ .$$

Let M^n be complete with $\mathrm{Ric}_{M^n} \geq 0$, $y \in M^n$, $r(x) = \rho_y(x)$. Then by 2) and 4) together with Proposition 7.7,

$$(\times) \qquad \Delta G(r(x)) \geq b$$

holds in the sense of barriers.

For f a Lipschitz function on M^n, denote by dil f the smallest constant k such that for all x_1, x_2,

$$|f(x_1) - f(x_2)| \leq k\, \overline{x_1, x_2}\ .$$

Theorem 7.9. *Let M^n be complete, with $\mathrm{Ric}_{M^n} \geq (n-1)H$. Let $u : B_{R+\eta}(y) \to R$ (for some $\eta > 0$) be a Lipschitz function satisfying*

i) $u \geq 0$,

ii) $u(y_0) = 0$,

for some $y_0 \in \overline{B_R(y)}$.

iii) dil $u \leq a$,

iv) $\Delta u \leq b$,

in the barrier sense. Then for all c, with $0 < c < R$

$$u(y) \leq a \cdot c + G(c)\ .$$

Proof: Take $\epsilon < \eta$ and define the function G using the value $R + \epsilon$, in place of R. Since we can eventually let $\epsilon > 0$, it will suffice to prove the inequality in this case.

In what follows we write G for $G(r(x))$. Fix $0 < c < R$ and suppose the bound is false. Then by iii), it follows that

$$u \mid \partial B_c(y) \geq G \mid \partial B_c(y)$$

Also, by i) and property 3) of the function G,

$$u \mid \partial B_{R+\epsilon}(y) \geq G \mid \partial B_{R+\epsilon}(y) \ .$$

Thus, the function $(G-u)$ satisfies

$$(G-u) \mid \partial B_c(y) \leq 0 \ ,$$

$$(G-u) \mid \partial B_{R+\epsilon}(y) \leq 0 \ .$$

However, by ii) and property 1) of G,

$$(G-u)(y_0) > 0 \ .$$

Hence, $(G-u) \mid \overline{B_{R+\epsilon}(y)} \setminus B_c(y)$ has a strict interior maximum. But since by iv) and $(\times)$ we know

$$\Delta(G-u) \geq 0$$

holds in the barrier sense, this contradicts the maximum principle (Theorem 7.8).

Remark 7.10. One easily checks that in the explicit formula for G in the case $H=0$, the optimal value of c is the unique number satisfying $0 < c < R$, and

$$c(\,(R/c)^n - 1) = \frac{an}{b} \ .$$

However, in Corollary 7.11 below, a value which is approximately optimal is all that is required.

As previously mentioned, Theorem 7.9 can be used to obtain an estimate on excess functions. Let

$$E(x) = \overline{x, y_1} + \overline{x, y_2} - \overline{y_1, y_2}$$

be the excess function associated to $y_1, y_2 \in M^n$. We can regard $E(x)$ as the excess of any triangle with vertices x, y_1, y_2 and all sides minimal.

Let γ be a minimal geodesic from y_1 to y_2. The function E satisfies

i) $E(x) \geq 0$,

ii) $E \mid \gamma \equiv 0$.

iii) dil $E \leq 2$.

In case $\mathrm{Ric}_{M^n} \geq 0$, by (++) above and Proposition 7.7, we have

iv) $\Delta E \leq (n-1)(\frac{1}{s_1} + \frac{1}{s_2})$,

where $s_j(x) = \overline{x, y_j}$.

Define the function $s(x)$ by

$$s(x) = \min(s_1(x), s_2(x)) \ .$$

Define the height function, $h(x)$, by

$$h(x) = \min_{\gamma, t} \overline{x, \gamma(t)} \ ,$$

where γ is a minimal segment from y_1 to y_2.

Clearly, $h(x) = 0$ implies $E(x) = 0$ and, by the triangle inequality,

$$E(x) \leq 2h(x) .$$

Under the assumption $\mathrm{Ric}_{M^n} \geq 0$, we now derive a quantitative relation between the values $h(x)$ and $s(x)$ which guarantees that $E(x)$ is small at the point x.

Corollary 7.11. *If* $\mathrm{Ric}_{M^n} \geq 0$, *then for* $h \leq \frac{1}{2}s$,

$$\begin{aligned} E &\leq 8\left(\frac{h^n}{s}\right)^{1/(n-1)} \\ &= 8\left(\frac{h}{s}\right)^{1/(n-1)} h . \end{aligned}$$

Proof: The function E satisfies the hypothesis of Theorem 7.9 with $a = 2$, $b = 4(n-1)/s$. Use $c = (2h^n/s)^{1/(n-1)}$ in the estimate

$$E \leq 2c + G(c) .$$

The sum of the first and third terms in the explicit expression for G is negative. The middle term gives a contribution at most equal to

$$\frac{2(n-1)}{sn} \cdot \frac{2}{n-2} h^n \left(\frac{2h^n}{s}\right)^{(2-n)/(n-1)} \leq 4c .$$

The claim follows immediately.

Remark 7.12. The estimate we have derived is of particular interest at points x, where h, s are large individually, but h^n/s is small. Roughly speaking, such triangles might be called "thin".

Remark 7.13. Suppose, in fact, that $K_{M^n} \geq 0$. Let y_0 be a closest point to x among all points which lie on minimal segments from y_1 to y_2. Divide a triangle with vertices x, y_1, y_2 into two right triangles with vertices x, y_1, y_0 and x, y_2, y_0. Put $\overline{y_0, y_j} = t_j$, $j = 1, 2$. Then Toponogov's theorem B) gives

$$(\surd) \qquad \begin{aligned} s_j &\leq (h^2 + t_j^2)^{1/2} , \\ &\leq t_j\left(1 + 2\left(\frac{h}{t_j}\right)^2\right) . \end{aligned}$$

Thus, for $t = \min(t_1, t_2)$, we have

$$E \leq \left(\frac{h}{t}\right) \cdot h .$$

If, at the point x, the values h, t are large but h^2/t is small, then E is still small. Note that such *thin* triangles are *not* required to be *as thin* as those in Remark 7.12.

Remark 7.14. Corollary 7.11 is the first estimate in which a nontrivial bound on a *sum of distances*, $s_1 + s_2$, is obtained from a bound on Ricci curvature. But at present, there is no useful bound on the individual s_j, as in $(\surd)$.

Appendix. The maximum principle.

Proof of Theorem 7.8: Let p be a weak local maximum i.e. $f(p) \geq f(x)$ for all x near p. Take a small normal coordinate ball, $B_\delta(p)$, and assume that there exists $z \in \partial B_\delta(p)$ such that $f(p) > f(z)$. Then, by continuity, $f(p) > f(z')$ for $z' \in \partial B_\delta(p)$ sufficiently close to z. Choose a normal coordinate system, $\{x_i\}$, such that $z = (\delta, 0, \cdots, 0)$. Put

$$\phi(x) = x_1 - d(x_2^2 + \cdots + x_n^2)\,,$$

where d is so large that if $y \in \partial B_\delta(p)$ and $f(y) = f(p)$, then $\phi(y) < 0$. Note that grad ϕ doesn't vanish.

Put

$$\psi = e^{a\phi} - 1\,.$$

Then

$$\Delta(\psi) = (a^2 \|\mathrm{grad}\ \phi\|^2 + a\,\Delta\phi)\,e^{a\phi}\,.$$

Thus, for a sufficiently large,

$$\Delta\psi > 0\,.$$

Moreover,

$$\psi(p) = 0\,.$$

For $\eta > 0$, sufficiently small,

$$(f + \eta\psi)|\partial B_\delta(p) < f(p)\,.$$

Thus, $f + \eta\psi$ has an interior maximum at some point $q \in B_\delta(p)$.

If $f_{q,\epsilon}$ is a barrier for f at q with $\Delta f_{q,\epsilon} \geq -\epsilon$, then $f_{q,\epsilon} + \eta\psi$ is also a barrier for $f + \eta\psi$ at q. For ϵ sufficiently small, we have

$$\Delta(f_{q,\epsilon} + \eta\psi) > 0\,.$$

Since $f + \eta\psi$ has a local maximum at q, and

$$f_{q,\epsilon} + \eta\psi \leq f + \eta\psi\,,$$

$$(f_{q,\epsilon} + \eta\psi)(q) = (f + \eta\psi)(q)\,,$$

we find that $f_{q,\epsilon} + \eta\psi$ has a local maximum at q as well. But this is incompatible with $\Delta(f_{q,\epsilon} + \eta\psi) > 0$. (Note that in normal coordinates at q, $\Delta = \partial_1^2 + \cdots + \partial_n^2$).

It follows that for all small δ, we have $f \mid \partial B_\delta(p) \equiv f(p)$. Since M is connected this implies $f \equiv f(p)$.

8. Nonnegative Ricci curvature, diameter growth and finiteness of topological type

In this section we prove that if M^n is complete, $\mathrm{Ric}_{M^n} \geq 0$, $K_{M^n} \geq -1$, and if a certain additional condition holds, then has finite topological type i.e. M^n is homeomorphic to the interior of a compact manifold with boundary.

The most general form of the additional condition uses the *ray density function*, $\mathcal{R}(r,p)$, associated to $p \in M^n$. However in some ways, a stronger condition formulated in terms of a second function, $\mathcal{D}(r,p)$, the *diameter growth function* is more natural.

Definition 8.1. Let M^n be complete. A *ray* is a geodesic, $\gamma : [0,\infty) \to M^n$, each segment of which is minimal.

When M^n is complete and noncompact then rays always exist.

Proposition 8.2. *Let M^n be complete noncompact. Then for all p, there exists at least one ray, γ, with $\gamma(0) = p$.*

Proof: Since M^n is not compact, there is a sequence, q_i, with $\overline{p,q_i} \to \infty$. Let γ_i be minimal from p to q_i and let $\{\gamma_j\}$ be a subsequence such that $\gamma_j'(0) \to v$, for some $v \in M_p^n$, with $\|v\| = 1$. Let $\gamma : [0,\infty) \to M^n$ be the geodesic with $\gamma'(0) = v$. Then each segment, $\gamma \mid [0,\ell]$, is a limit of minimal segments, $\gamma_j \mid [0,\ell]$, and hence is minimal itself. Thus γ is a ray.

Let $x \in M^n$, and let γ be a ray from p. Note that if

$$t_1 \leq \overline{x,p} - \overline{x,\gamma}$$

$$t_2 \geq \overline{x,p} + \overline{x,\gamma}$$

then

$$\overline{x,\gamma} = h(x) \,,$$

where $h(x)$ is the height function of §7, for the excess function associated to the points $\gamma(t_1)$, $\gamma(t_2)$.

Definition 8.3. Define the ray density function by

$$\mathcal{R}(r,p) = \sup_{x \in \partial B_r(p)} \{\inf_{\gamma} \overline{x,\gamma} \mid \gamma \text{ a ray, } \gamma(0) = p\} \,.$$

Proposition 8.4. *Let M^n be complete, with* $\mathrm{Ric}_{M^n} \geq 0$ *on $M^n \setminus B_\lambda(p)$, for some $\lambda < \infty$. Assume*

$$\mathcal{R}(r,p) = o(r^{1/n}) \,.$$

Then for all $\epsilon > 0$ there exists $\delta > 0$, such that if x, γ are as in Definition 8.3 (x arbitrary) with

$$\overline{x,p} \geq \delta^{-1} \,,$$

and t is sufficiently large relative to $\overline{x,p}$, then the excess function associated to $\gamma(0)$, $\gamma(t)$ satisfies

$$E(x) \leq \epsilon \,.$$

Proof: For the case in which $\mathrm{Ric}_{M^n} \geq 0$ on all of M^n, this is immediate from Corollary 7.11. The general case is dealt with in the Appendix to this section.

On the other hand, in case $K_M \geq -1$, the following proposition provides a positive lower bound for $E(x)$ when x is critical with respect to say y_1.

Put $\overline{x, y_1} = s$.

Proposition 8.5. *Let M^n be complete with $K_{M^n} \geq -1$ $(H < 0)$. Let x be critical with respect to y_1. Then for all $\epsilon > 0$ there exists δ such that*

$$\overline{x, y_2} \geq \delta^{-1}$$

implies

$$E(x) \geq \ell n\Big(\frac{2}{1+e^{-2s}}\Big) - \epsilon\ .$$

Proof: By Toponogov's theorem B) and the assumption that x is critical with respect to y_1, it suffices to assume that $x, y_1, y_2 \in M^2_{-1} \subset M^n_{-1}$ and that the minimal geodesics from x to y_1, y_2 make an angle $\pi/2$. By hyperbolic trigonometry

$$\cosh \overline{y_1, y_2} = \cosh s \cosh \overline{x, y_2}$$

By the triangle inequality,

$$|\overline{y_1, y_2} - \overline{x, y_2}| \leq s\ .$$

As both $\overline{y_1, y_2}$, and $\overline{x, y_2} \to \infty$ with s fixed,

$$\frac{\cosh \overline{x, y_1}}{\cosh \overline{y_1, y_2}} \quad \to \quad e^{\overline{x,y_1} - \overline{y_1,y_2}}$$

The claim follows easily.

By combining Propositions 8.4 and 8.5 we obtain

Theorem 8.6 (Abresch-Gromoll). *Let M^n be complete with*

i) $\mathrm{Ric}_{M^n} \geq 0$ *on* $M^n \setminus B_\lambda(p)$, *for some* λ,

ii) $\mathcal{R}(r,p) = o(r^{1/n})$, *for some* $p \in M^n$,

iii) $K_{M^n} \geq H > -\infty$.

Then there exists a compact set, C, such that $M^n \backslash C$ contains no critical points of p. In particular, M^n has finite topological type.

We now define the diameter growth functin $\mathcal{D}(r,p)$. For every r, the open set $M^n \setminus \overline{B_r(p)}$ contains only *finitely many unbounded components*, U_r. Each U_r has *finitely many* boundary components, $\Sigma_r \subset \partial B_r(p)$. In particular Σ_r is a closed subset.

Let $\mathrm{dia}(\Sigma_r)$ denote maximum distance, *measured in M^n*, between a pair of points of Σ_r.

Definition 8.7.

$$\mathcal{D}(r,p) = \sup_{\Sigma_r} \mathrm{dia}(\Sigma_r)\ .$$

Given any boundary component, Σ_r, we can construct a ray, γ, such that $\gamma(t) \subset U_r$, for $t > r$ and so, $\gamma(r) \in U_r$. To do so, it suffices to choose the sequence of points, $\{q_i\}$, of Proposition 8.3 to lie in U_r. Then the convergent subsequence γ_j satisfies the conditions above. Hence γ satisfies them as well.

With this observation, it follows immediately from the proof of Theorem 8.6, that if we assume

$$\mathcal{D}(r,p) = o(r^{1/n}) ,$$

then for $r \geq r_0$ sufficiently large, no point of any set Σ_r is critical with respect to p.

Fix r_0, U_{r_0}, a boundary component Σ_{r_0} and a ray γ, with $\gamma(r_0) \in \Sigma_{r_0}$, $\gamma(t) \in U_{r_0}$, for $t > r_0$. For each $t \geq r_0$, let Σ_t denote the boundary component of the unbounded component of $M^n \setminus \overline{B_t(p)}$ with $\gamma(t) \in \Sigma_t$. Using the observation of the previous paragraph and the Isotopy Lemma 1.4, we easily construct an imbedding,

$$\psi : (r_0, \infty) \times \Sigma_0 \to U_x ,$$

such that

$$\psi((t, \Sigma_{r_0})) = \Sigma_t .$$

It follows easily that $\psi((r_0, \infty) \times \Sigma_{r_0}))$ is open and closed in U_{r_0}. Hence

$$\psi((r_0, \infty) \times \Sigma_{r_0}) = U_{r_0} .$$

Thus we obtain

Theorem 8.8 (Abresch-Gromoll). *Let M^n be complete with*

i) $\mathrm{Ric}_{M^n} \geq 0$ *on* $M^n \setminus B_\lambda(p)$, *for some* λ,

ii) $\mathcal{D}(r,p) = o(r^{1/n})$ *for some* $p \in M^n$,

iii) $K_{M^n} \geq -H > -\infty$.

Then

$$\mathcal{R}(r,p) = o(r^{1/n}) .$$

Thus there exists a compact set C such that $M^n \setminus C$ contains no critical points of p. In particular, M^n has finite topological type.

Remark 8.9. Clearly, any two points on $\partial B_r(p)$ can be joined by a broken geodesic passing through p of length $2r$. Thus, one always has

$$\mathcal{D}(r,p) \leq 2r ,$$

for the function, $\mathcal{D}(r,p)$, defined in Definition 8.7. However, it is also of interest to consider modified definitions of diameter growth, for which the above inequality, need not hold. For such definitions and their geometric significance, see [AGl], [Liu], [Shen], [Z].

Remark 8.10. Examples of and [ShY] show that if i) and iii) of Theorem 8.8 are retained but (e.g. if $n = 7$) ii) is weakened to $\mathcal{D}(r,p) = O(r^{1/2})$ then the conclusion fails; see also [AnKLe].

Remark 8.11. For further results related to those of this section; see Shen.

Appendix. Nonnegative Ricci curvature outside a compact set.

Proof of Proposition 8.4: The triangle inequality implies that for $0 < t_1 < t$ the excess function, $\widetilde{E}$, associated to $\gamma(t_1)$, $\gamma(t)$ satisfies

$$E \leq \widetilde{E}$$

(where E is the excess function associated to $\gamma(0)$, $\gamma(t)$). Thus it suffices to show that if $\overline{x,p}$ is sufficiently large and t_1, t are suitably chosen, then $\widetilde{E}$ can be made arbitrarily small.

Clearly, we need only ensure that in the present more general situation the bounds on $\Delta G, \Delta\widetilde{E} \mid B_{h(x)}(x)$ are just as in the case in which $\mathrm{Ric}_{M^n} \geq 0$ on all of M^n. This is clear for ΔG, since $B_{h(x)}(x) \cap B_\lambda(p) = \emptyset$, provided $\overline{x,p}$ is sufficiently large.

As for the function $\widetilde{E}$, it clearly suffices to know that a minimal geodesic, σ, from $\gamma(t_1)$ or $\gamma(t)$ to $z \in B_{h(x)}(x)$ does not intersect $B_\lambda(p)$. Consider, for definiteness, the point $\gamma(t_1)$ and suppose $\sigma \cap B_\lambda(p) \neq \emptyset$. Then by the triangle inequality,

$$(*) \qquad \overline{\gamma(t_1), z} > (t_1 - \lambda) + (\overline{x,p} - h(x) - \lambda) ,$$

On the other hand, if $\underline{t}$ is a point on γ closest to x, clearly

$$\overline{\gamma(\underline{t}), z} \leq 2h(x) .$$

Also

$$\underline{t} - t_1 \leq \overline{x,p} + h(x) - t_1 .$$

Combining these gives

$$(**) \qquad \overline{\gamma(t_1), z} < \overline{x,p} + 3h(x) - t_1 .$$

From $(*)$, $(**)$, we get

$$t_1 < 2h(x) + \lambda ,$$

and, we can take $t_1 = 2h(x) + \lambda$.

The argument for $\gamma(t)$ is similar to the one just given.

References

[A] U. Abresch: Lower curvature bounds, Toponogov's theorem, and bounded topology I, II, Ann. scient. Ex. Norm. Sup. 18 (1985) 651-670 (4e série), and preprint MPI/SFB 84/41 & 45.

[AG1] U. Abresch and D. Gromoll, "On complete manifolds with nonnegative Ricci curvature, J.A.M.S. (to appear).

[An] M. Anderson, Short geodesics and gravitational instantons, J. Diff. Geom. V. 31 (1990) 265-275.

[AnKle] M. Anderson, P. Kronheimer and C. LeBrun, Complete Ricci-flat Kahler manifolds of inifnite topological type, Commun. Math.Phys. 125 (1989) 637-642.

[Be] M. Berger, Les varietés $\frac{1}{4}$-pincées, Ann. Scuola Norm. Pisa (111) 153 (1960) 161-170.

[BT] R. Bott and L. Tu, Differential forms in algebraic topology, Graduate Texts in Math., Springer-Verlag (1986).

[Ca] E. Calabi, An extension of E. Hopf's maximum principle with an application to Riemannian geometry, Duke Math. J. 25 (1958) 45–56.

[C1] J. Cheeger, Comparison and finiteness theorems in Riemannian geometry, Ph.D. Thesis, Princeton Univ., 1967.

[C2] ———, The relation between the Laplacian and the diameter for manifolds of nonnegative curvature, Arch. der Math. 19 (1968) 558-560.

[C3] ———, Finiteness theorems for Riemannian manifolds, Amer. J. Math. 92 (1970) 61-74.

[CE] J. Cheeger and D. G. Ebin, Comparison Theorems in Riemannian Geometry, North-Holland Mathematical Library 9 (1975).

[CGl1] J. Cheeger and D. Gromoll, The splitting theorem for manifolds of nonnegative Ricci curvature, J. Diff. Geo. 6 (1971) 119-128.

[CGl2] ———, On the structure of complete manifolds of nonnegative curvature, Ann. of Math. 96 (1972) 413-443.

[EH] J. Eschenburg and E. Heintze, An elementary proof of the Cheeger-Gromoll splitting theorem, Ann. Glob. Anal. & Geom. 2 (1984) 141-151.

[GrP] K. Grove and P. Petersen, Bounding homotopy types by geometry, Ann. of Math. 128 (1988) 195-206.

[GrPW] K. Grove, P. Petersen, J. Y. Wu, Controlled topology in geometry, Bull. AMS 20 (2) (1989) 181-

[GrS] K. Grove and K. Shiohama, A generalized sphere theorem, Ann. of Math. 106 (1977) 201-211.

[G] M. Gromov, Curvature, diameter and Betti numbers, Comm. Math. Helv. 56 (1981) 179-195.

[GLP] M. Gromov, J. Lafontaine, and P. Pansu, Structure Métrique pour les Variétiés Riemanniennes, Cedic/Fernand Nathan (1981).

[GreWu] R. Greene and H. Wu, Lipschitz convergence of Riemannian manifolds, Pacific J. Math. 131 (1988) 119-141.

[Liu] Z. Liu, Ball covering on manifolds with nonnegative Ricci curvature near infinity, Proc. A.M.S. (to appear).

[M] J. Milnor, Morse Theory, Annals of Math. Studies, Princeton Univ. Press (1963).

[Me] W. Meyer, Toponogov's Theorem and Applications, Lecture Notes, Trieste,1989.

[Pe1] S. Peters, Cheeger's finiteness theorem for diffeomorphism classes of Riemannian manifolds, J. Reine Angew. Math. 394 (1984) 77-82.

[Pe2] ———-, Convergence of Riemannian manifolds, Comp. Math. 62 (1987) 3-16.

[ShY] J. P. Sha and D. G. Yang, Examples of manifolds of positive Ricci curvature, J. Diff. Geo. 29 (1) (1989) 95-104.

[Shen] Z. Shen, Finite topological type and vanishing theorems for Riemannian manifolds, Ph.D. Thesis, SUNY, Stony Brook, 1990.

[We] A. Weinstein, On the homotopy type of positively pinched manifolds, Archiv. der Math. 18 (1967) 523-524.

[Z] S. Zhu, Bounding topology by Ricci curvature in dimension three, Ph.D. Thesis, SUNY, Stony Brook, 1990.

Rigidity of lattices : An introduction

by M. GROMOV and P. PANSU

We present in these lectures basic facts and ideas in the geometry of discrete subgroups in Lie groups with a special emphasis laid upon the rigidity of lattices in the semi-simple Lie groups. We try to give a broad panorama of the field and explain various approaches to the study of the rigidity. These can be roughly divided into two categories. The first approach developed by G.D. Mostow [1967] and H. Fürstenberg [1967] (see $[\mathrm{Mos}]_1$ and $[\mathrm{Für}]_1$) uses the geometry and dynamics of the action of discrete subgroups $\Gamma \subset G$ on an appropriate *ideal boundary* of the ambient Lie group G. This geometro-dynamical line of development culminated in the *superrigidity and arithmeticity* theorems for lattices in the simple Lie groups of $\mathbb{R}$-rank ≥ 2, proven by G.A. Margulis in 1974 (see $[\mathrm{Mar}]_1$).

The second approach, more analytic in nature, uses elliptic P.D.E. and Bochner type integro-differential inequalities. The first results here obtained by E. Calabi and E. Vesentini [1960] and then by A. Weil [1962] (see [Cal], [Cal-Ves], [Wei]), dealt with local and infinitesimal deformations of $\Gamma \subset G$ and the P.D.E's involved were linear. The Calabi-Vesentini method was delinearized by Y.T. Siu in 1980 who found a Kodaira-Bochner-(Siu) identity for harmonic maps of Kähler manifolds into Riemannian manifolds satisfying certain negative curvature conditions. Using Siu's method, N. Mok [1989] was able to give an alternative proof of Margulis' superrigidity for the (lattices in the isometry groups G of) Hermitian symmetric spaces. More recently, K. Corlette [1990] has found a Bochner formula in the quaternionic case and thus established the superrigidy for discrete groups of isometries of the quaternionic hyperbolic spaces as well as for the Cayley plane (see $[\mathrm{Cor}]_1$). Notice that these cases are not covered by Margulis' theory.

Our presentation of both methods, geometro-dynamic and analytic, is quite superficial as we try to avoid all difficult spots as much as possible. Yet we provide the proofs of the basic elementary facts which convey the flavor of the deeper aspects of the theory.

We conclude this introduction with our thanks to Franco Tricerri and Paolo de Bartolomeis who brought us to the comfort of Montecatini where we could calmly present our lectures.

1. Generalities on Lie groups and discrete subgroups.

There are two sharply distinct classes of closed subgroups H in a connected Lie group G. The first case is where H is connected. Then by the classical theory (see [Pont]) H is a Lie (sub)-group which is uniquely determined by its Lie subalgebra $L(H) \subset L(G)$. Thus all geometric properties of the subgroup $H \subset G$ are encoded in the linear algebra of $L(H) \subset L(G)$. The decoding may be difficult at times, but, in principle, it is always possible.

The opposite case is where the subgroup is discrete in G. Here we prefer to use the notation Γ rather than H as this Γ is an animal of quite different nature from G.

The linear structure of $L(G)$ can not be used in the study of discrete subgroups $\Gamma \subset G$ as the infinitesimal information contained in $L(\Gamma)$ literally reduces to zero. Yet as we shall see in these lectures one can use global geometric methods and achieve a fair understanding of Γ in many cases.

1.1. Elementary example. Before going into any kind of general theory we want to revive in the reader's mind the familiar picture of a lattice in the plane. We take $G = \mathbb{R}^2$ and consider the subgroup $\Gamma \subset \mathbb{R}^2$ consisting of all integral combinations of two linearly independent vectors x and y in $\mathbb{R}^2$, that is

$$\Gamma = \{mx + ny\}_{m,n \in \mathbb{Z}} \subset \mathbb{R}^2 .$$

This Γ is obviously discrete and as an abstract group it is isomorphic to $\mathbb{Z}^2$. The quotient space $\mathbb{R}^2/\Gamma$ is compact. In fact, this quotient is, as everybody knows, the 2-torus.

There is another kind of discrete subgroup $\Gamma \subset \mathbb{R}^2$ where the quotient is non-compact. Namely we may take Γ consisting of the multiples of a single non-zero vector in $\mathbb{R}^2$,

$$\Gamma = \{mx\}_{m \in \mathbb{Z}} \subset \mathbb{R}^2 .$$

Here the quotient space is the infinite cylinder $S^1 \times \mathbb{R}^1$.

1.2. Definitions. A (discrete) subgroup $\Gamma \subset G$ is called *cocompact* if the quotient space G/Γ is compact. An equivalent condition is the existence of a *compact* subset $D \subset G$ whose Γ-translates cover G. That is

$$\bigcup_{\gamma \in \Gamma} \gamma D = G .$$

Yet another way to define the cocompactness is to require that Γ is *a net* in G for some (and hence for every) left invariant Riemannian metric in G, where a subset Γ in G (with a given metric) is called a net if there exists (possibly large) $\varepsilon \geq 0$, such that for every $g \in G$ there exists some $\gamma \in \Gamma$ such that $\operatorname{dist}(g,\gamma) \leq \varepsilon$. In other words, the *$\varepsilon$-neighbourhood*

$$U_\varepsilon(\Gamma) \underset{\text{def}}{=} \{g \in G \mid \operatorname{dist}(g,\Gamma) \leq \varepsilon\}$$

equals all of G.

The equivalence of the three definitions is rather obvious. What distinguishes the first definition is the appeal to the quotient space G/Γ (which perversely is sometimes denoted by $\Gamma\backslash G$) which a priori carries less information than the pair $(G, \Gamma \subset G)$. The last definition with nets is interesting as it makes sense for arbitrary metric spaces (where one usually specifies an ε and speaks of ε-nets).

An underlying intuitive idea of the above definition is that cocompact subgroups $\Gamma \subset G$ provide a fair discrete approximation of G with a bounded error and so their properties are expected to be similar to those of G. A more serious mathematical reason for introducing the notion of cocompactness is the existence of certain remarkable subgroups with this property as indicated in 1.4.B.

Our next definition also expresses the idea of approximation of G by Γ but now in terms of measure theory.

A discrete subroup $\Gamma \subset G$ is called a *lattice* (or a *finite covolume* subgroup) if G/Γ has *finite* volume, i.e. the measure on G/Γ induced by the Haar measure on G has finite total mass. An equivalent property is the existence of a non-trivial finite G-invariant measure on G/Γ. The third definition is the existence of a subset $D \subset G$ of finite Haar measure whose Γ-translates cover G. (In these definitions, if we speak of a left invariant Haar measure, we must use the left action of Γ on G. Or we could go to the right measure and action. Fortunately, the existence of a single lattice $\Gamma \subset G$ implies by an easy argument that every left invariant measure on G/Γ is right invariant which makes the left-right precautions unnecessary).

1.3. Simple example. Let us try the above definition on discrete subgroups Γ of the group $G = \mathbb{R}^n$. Every such Γ equals the integral span of some linearly independent vectors $x_1,\dots,x_k$ in $\mathbb{R}^n$, as an easy (and well known) argument shows. Then the quotient $\mathbb{R}^n/\Gamma$ is the Cartesian product of the torus $T^k = \mathbb{R}^k/\Gamma$ by $\mathbb{R}^{n-k}$ where $\mathbb{R}^k \subset \mathbb{R}^n$ is the span of $x_1,\dots,x_k$ and $\mathbb{R}^{n-k} = \mathbb{R}^n/\mathbb{R}^k$. Now one sees that the following four conditions are equivalent:

1. Γ is cocompact in $\mathbb{R}^n$.

2. Γ has finite covolume.

3. Γ is isomorphic, as an abstract group, to $\mathbb{Z}^n$.

4. The linear span of Γ in $\mathbb{R}^n$ equals $\mathbb{R}^n$.

1.4. Non-cocompact lattices. The above example may make one believe that

$$\text{cocompact} \Leftrightarrow \text{finite covolume.}$$

In fact the implication

$$\text{cocompact} \Rightarrow \text{finite covolume}$$

is true and obvious as every *compact* subset $D \subset G$ has finite haar measure. The opposite implication is known to be true for nilpotent and solvable Lie groups (see [Rag]) as well as for $G = \mathbb{R}^n$ treated above. Yet this is not so for *semisimple* Lie groups and the counter example is provided by the following

1.4.A. Most remarkable lattice. Take $G = SL_n\mathbb{R}$ and let $\Gamma = SL_n\mathbb{Z} \subset SL_n\mathbb{R}$ that is the subgroup consisting of the matrices with integral entries and having determinant one. For example, if $n = 2$, then $\Gamma = SL_2\mathbb{Z}$ equals the set of the *integral* solutions a, b, c, d to the equation $ab - cd = 1$.

1.4.A'. Theorem. $SL_n\mathbb{Z}$ *has finite covolume in* $SL_n\mathbb{R}$ *for* $n \geq 2$ *but is not cocompact.*

The proof of the finite covolume property follows from the Hermite-Minkowski reduction theory which provides an "almost orthogonal"basis in every lattice in $\mathbb{R}^n$ (see [Rag], [Cas]).

Notice that the finite covolume property of $\Gamma \subset G$ implies Γ is infinite in the (interesting) case where G is non-compact as $\infty = \mathrm{Vol}\, G = (\#\,\Gamma)\, \mathrm{Vol}(G/\Gamma)$. For example the Diophantine equation $ab - cd = 1$ has infinitely many solutions. This can also be seen directly by successively multiplying the matrices $\begin{pmatrix} 1 & 0 \\ 1 & 1 \end{pmatrix}$ and $\begin{pmatrix} 1 & 1 \\ 0 & 1 \end{pmatrix}$ in $SL_2\mathbb{Z}$.

1.4.A". Geometric interpretation of the quotient space $SL_n\mathbb{R}/SL_n\mathbb{Z}$. We claim that $GL_n\mathbb{R}/SL_n\mathbb{Z}$ equals the space of unimodular lattices in $\mathbb{R}^n$. To see that we observe the following two facts

(a) $SL_n\mathbb{Z}$ consists of exactly those linear maps $\mathbb{R}^n \to \mathbb{R}^n$ with determinant one which send the subgroup $\mathbb{Z}^n \subset \mathbb{R}^n$ (necessarily bijectively) into itself.

(b) Let $\mathfrak{L} \subset \mathbb{R}^n$ be a *unimodular* lattice in $\mathbb{R}^n$, that is $\mathrm{Vol}\,\mathbb{R}^n/\mathfrak{L} = 1$ for the standard Haar (Lebesgue) measure on $\mathbb{R}^n$. Then there exists a linear transformation g of $\mathbb{R}^n$ with $\det g = 1$ which maps $\mathfrak{L}$ onto $\mathbb{Z}^n$.

Proof. As we have mentioned earlier, $\mathfrak{L}$ is integrally spanned by some vectors $x_1,\ldots,x_n$ in $\mathbb{R}^n$ (one may use any $x_1,\ldots,x_n$ which generate $\mathfrak{L}$ as a group). Then we use a transformation g which send $x_1,\ldots,x_n$ to the standard basis of $\mathbb{R}^n$ (which integrally generates $\mathbb{Z}^n$).

Now we see that the action of $SL_n\mathbb{R}$ on the space $\{\mathfrak{L}\}$ of the unimodular lattices in $\mathbb{R}^n$ is transitive and according to (a) the isotropy group at $\mathfrak{L} = \mathbb{Z}^n$ equals $SL_n\mathbb{Z}$. Thus

$$\{\mathfrak{L}\} = SL_n\mathbb{R}/SL_n\mathbb{Z},$$

as we have stated.

Now, to show that $SL_n\mathbb{R}/SL_n\mathbb{Z}$ is non-compact, we take a sequence of unimodular lattices $\mathfrak{L}_i$, $i = 1,2,\ldots$, where $\mathfrak{L}_i$ is generated by vectors $x_1^i, x_2^i,\ldots x_n^i$, such that $\|x_1^i\| \to 0$ as $i \to \infty$. Clearly, such a sequence diverges in $SL_n\mathbb{R}/SL_n\mathbb{Z}$ with the quotient space topology.

Remark. The effect of the condition $\|x_1^i\| \to 0$ can best be seen in the geometry of the torus $T_i^n = \mathbb{R}^n/\mathfrak{Z}_i$. Namely, the geodesic loop in T_i^n covered by the segment $[0, x_1^i] \subset \mathbb{R}^n$ has length $= \|x_1^i\| \to 0$ and so Inj Rad $T_i^n \to 0$ for $i \to \infty$. On the other hand, the convergence $\mathfrak{Z}_i \to \mathfrak{Z}$ would imply a convergence (see below) of T_i^n to some flat n-dimensional torus T^n. Such a torus, of course, has Inj Rad $T^n = \rho > 0$ which contradicts to the convergence Inj Rad $T_i^n \to 0$ as Inj Rad is a *continuous* function on the space of flat tori, with the following topology corresponding to that in $SL_n\mathbb{R}/SL_n\mathbb{Z}$: a sequence T_i^n converges to T^n if and only if there exist linear diffeomorphisms $f_i : T_i^n \to T^n$, such that the induced metrics $f_i^*(g)$ converge to g_i, where g and g_i are the flat Riemannian metrics on T^n and T_i^n coming from the Euclidean metric on $\mathbb{R}^n$.

1.4.B. Arithmetic lattices. Starting from $SL_n\mathbb{Z}$ in $SL_n\mathbb{R}$ on may construct further (arithmetic) lattices in Lie groups. The construction goes in three steps.

(1) A *lattice* $\Gamma \subset G$ is called 1-*arithmetic* if there exists a continuous homomorphism h_1 of G into $GL_n\mathbb{R}$ for some n, such that $\Gamma = h_1^{-1}(SL_n\mathbb{Z})$.

(2) A lattice $\Gamma \subset G$ is called 2-*arithmetic* if there exists a 1-arithmetic lattice Γ_1 in some Lie group G_1 for which there exists a continuous homomorphism $h_2 : G_1 \to G$ with *compact* kernel, such that $h_2(\Gamma_1) = \Gamma$.

(3) A lattice $\Gamma \subset G$ is called *arithmetic* if it is *commensurable* with a 2-arithmetic lattice $\Gamma_2 \subset G$, where the commensurability means that the intersection $\Gamma \cap \Gamma_2$ has a finite index in Γ as well as in Γ_2.

Remarks. (1) Our definition of arithmeticity is slightly different from the usual one (see $[\mathrm{Bor}]_1$ [Rag]) but it is equivalent to the standard definition for semisimple groups G.

(2) If we want to use this construction to produce an (arithmetic) lattice in G we may first enlarge G by (essentially) multiplying it by a compact group K and then we need a homomorphism h_1 of the enlarged group $G_1 = G \times K$ into some $GL_n\mathbb{R}$, such that h_1^{-1} is a *lattice* in G_1. Constructing such G_1 and then h_1 is by no means trivial. However, A. Borel (see $[\mathrm{Bor}]_2$) has proven by this method that every non-compact simple Lie group G contains some arithmetic cocompact lattice and also some non-cocompact lattice.

Example. Let

$$G = O(\varphi_0) = O(p,q) \subset GL_n \ , \ n = p+q \ ,$$

be the orthogonal group consisting of the linear transformations of $\mathbb{R}^k$ fixing the form

$$\varphi_0 = \sum_{i=1}^{p} x_i^2 - \sum_{i=p+1}^{n} x_i^2 .$$

Then we take another form, say φ_1 of the same signature as φ_0 say

$$\varphi_1 = \sum_{i=1}^{p} a_i x_i^2 - \sum_{i=p+1}^{n} b_i x_i^2 ,$$

for $a_i > 0$ and $b_i > 0$, and denote by $O(\varphi_1)$ the orthogonal group corresponding to φ_1. The two forms are linearly equivalent and the linear transformation $g \in GL_n$ which moves φ_0 to φ_1 also moves $O(\varphi_0)$ to $O(\varphi_1)$, by

$$gO(\varphi_0)g^{-1} = O(\varphi_1) .$$

Thus we obtain with φ_1 a new embedding to GL_n with the image $O(\varphi_1)$, say $h_1 : O(p,q) \to GL_n$. Now we ask ourselves when $h_1^{-1}(SL_n\mathbb{Z})$ (which is the same as $O(\varphi_1) \cap (SL_n\mathbb{Z})$) is a lattice in $O(p,q)$. The answer is provided by the following

Theorem. *Let* $n \geq 3$ *and both numbers* p *and* q *be positive.* (If one of them is zero the group $O(\varphi)$ is compact and every $\Gamma \subset O(\varphi)$ is a cocompact lattice). *Then* $\Gamma = SL_n\mathbb{Z} \cap O(\varphi_1)$ *is a lattice in* $O(\varphi_1)$ *if and only if* φ_1 *is proportional to a rational form,* $\varphi_1 = \alpha\varphi'_1$ for

$$\varphi'_1 = \sum_{i=1}^{p} a'_i x_i^2 - \sum_{i=p+1}^{n} b'_i x_i^2$$

where a'_i *and* b'_i *are rational numbers. Furthermore,* Γ *is cocompact if the only rational solution to the equation*

$$\varphi'_1(x_1,x_2...x_n) = 0$$

is

$$x_1 = x_2 = ... = x_n = 0 .$$

The proof follows from the general theory of arithmetic groups and can be found in [Bor]$_2$ and [G-P].

Remarks. (a) The theorem implies, in particular, that the group $\Gamma = SL_n\mathbb{Z} \cap O(\varphi_1)$ is *infinite* for all rational indefinite forms φ_1.

(b) The theorem remains valid for $n = 2$ unless the form φ'_1 splits into the product of two rational linear forms. For example Γ is a (cocompact) lattice for $\varphi = x_1^2 - bx_2^2$, where b is a positive rational number which is *not* a square of a rational number. Now, look at the orbit $\gamma(e_1) \subset \mathbb{R}^2$ for $e_1 = (1,0) \in$

$\mathbb{R}^2$ and all $\gamma \in \Gamma$. As Γ is infinite this orbit is also infinite. On the other hand, every $\gamma(e_1)$ has *integer* coordinates x_1 and x_2 such that

$$x_1^2 - bx_2^2 = \varphi(\gamma(e_1)) = \varphi(e_1) = 1 .$$

Thus we recapture the classical result on the infinity of integral solutions to the Pell equation $x_1^2 - bx_2^2 = 1$.

(c) If $n \geq 5$, then the equation $\varphi'_1(x_1,...,x_n) = 0$ always has a non-trivial rational solution (see [B-S]) and so our $\Gamma \subset O(\varphi)$ is not cocompact. Yet one can obtain cocompact lattices by using quadratic forms over *finite extensions* of $\mathbb{Q}$ and invoking the second step of the definition (see 2.9.C_2, $[\mathrm{Bor}]_1$, [G-P]).

1.5. Discrete groups and locally homogeneous spaces. Let the Lie group G in question equal the isometry group of a simply connected homogeneous Riemannian manifold. Then the structure of every discrete subgroup $\Gamma \subset G = \mathrm{Iso}\, X$ is adequately reflected in the geometry of the quotient space $V = X/\Gamma$. For example, if Γ has *no torsion* and thus the action of Γ on X is free, then X equals the *universal covering* of V and Γ appears as the deck transformation (Galois) group of the covering $X \to V$. (In fact this remains true in the torsion case if V is given the quotient *orbifold* structure).

1.5.A. *Basic Example.* Let X be a symmetric space of negative Ricci curvature. Then X necessarily is simply connected and has non-positive sectional curvature (see [B-G-S]). Thus X is homeomorphic to a Euclidean space. Furthermore, the isometry group $G = \mathrm{Iso}\, X$ is semisimple and $X = G/K$ for the maximal compact subgroup $K \subset G$. Now, one sees that the compactness of K implies that

$$\text{compactness } G/\Gamma \Leftrightarrow \text{compactness } X/\Gamma$$

$$\mathrm{Vol}\, X/\Gamma < \infty \Leftrightarrow \mathrm{Vol}\, G/\Gamma < \infty .$$

Hence, the study of torsion free cocompact discrete subgroups $\Gamma \subset G$ reduces to the geometry of compact locally symmetric spaces V locally isometric to X. Similarly, lattices $\Gamma \subset G$ correspond to complete manifolds V of finite volume.

Remarks. (a) The torsion free condition is not very restrictive for the following two reasons. First, every lattice Γ in our G contains a torsion free subgroup $\Gamma' \subset \Gamma$ of finite index (see [Rag]) and many questions concerning Γ can be easily readdressed to Γ'. However, the existence of Γ' is a highly non-trivial matter and, in fact, one can easily do without it by slightly enlarging the category. Namely, one should allow the *singular* spaces $V = X/\Gamma$, where Γ may have torsion. These can be treated as *orbifolds* (see [Th]) or as *metric* spaces, where the Riemannian metric X defines a curve length and consequently a metric in V. Notice that V contains a well defined nonsingular locus $U \subset V$ which is open and dense in V and which is characterized by the following property. The lift Y of U to X (for the quotient map $X \to V$) is the maximal subset in X on which Γ acts *freely*. Observe that this U is a non-complete Riemannian manifold which is locally isometric to X and the metric completion of U equals V.

(b) Not every semisimple group G appears as $\mathrm{Iso}\, X$ for some symmetric space X. For example, the universal covering of $SL_2\mathbb{R}$ is not like that. However, every G is *locally isomorphic* to $G' = \mathrm{Iso}\, X$

for some X and the local isomorphism (i.e. the isomorphism between the universal coverings) is as good for our problems as an actual isomorphism.

Summarizing (a) and (b) we come to the conclusion that one can see much of discrete groups Γ in semisimple Lie groups G by looking at complete locally symmetric spaces V. Conversely, the geometry of these V reduces to that of discrete groups Γ isometrically acting on the universal coverings X of V.

Although the two view points are essentially equivalent they bring along quite different geometric images and techniques. When we look at a lattice Γ in a non-compact Lie group G we see a periodic set of stars in an infinite sky. Mathematically one thinks of Γ as a kind of a discrete approximation to G and (or) regards G as a kind of a continuous envelope (or hull) of Γ. But nothing of this is seen in $V = X/\Gamma$. This is just a complete manifold of finite volume (which is compact if $\Gamma \subset G$ is cocompact). The study of such V can be naturally conducted in the general framework of Riemannian geometry and analysis on (compact) manifolds.

§2. An overview of the rigidity problems.

The word "rigidity" applies to (discrete) subgroups $\Gamma \subset G$ as well as to locally homogeneous manifolds (and orbifolds) V. This expresses the idea that Γ is "unmovable" in G and in terms of V that the topology of V determines the (locally homogeneous) geometry. One usually distinguishes three different kinds of rigidity.

I. *Ordinary* or *local rigidity* which is sometimes called just *rigidity.* (See 2.1.).

II. *Strong (Mostow) rigidity.* (See 2.2.A.)

III. *Superrigidity* discovered by Margulis. This has two aspects, Archimedean and non-Archimedean (see 2.9. and 2.10.).

2.1. Rigidity of homomorphisms. Let Γ be an abstract countable group and G be a Lie group. A homomorphism $\rho : \Gamma \to G$ is called (locally) *rigid* if every deformation of ρ is induced by automorphisms of the ambient group G. Let us explain it in more details. First, a *one parameter deformation* ρ_t, $t \in \mathbb{R}$, is a family of homomorphisms such that $\rho_0 = \rho$ and $\rho_t(\gamma) \in G$ is continuous in t for every $\gamma \in \Gamma$.

There is an obvious way to deform any $\rho = \rho_0$. Just take a path in G, that is $g_t \in G, t \in \mathbb{R}$, with $g_0 = \mathrm{id}$ and compose ρ_0 with the conjugations by g_t,

$$\rho_t = g_t \rho g_t^{-1} . \qquad (*)$$

More generally, one can take a one parameter family of automorphisms a_t of G with $a_0 = \mathrm{Id}$ and then deform ρ by

$$\rho_t = a_t \circ \rho . \qquad (**)$$

The deformations of this kind (**) have little to do with the specific of Γ and ρ and they are called *trivial.* (Sometimes one reserves the word "trivial" to inner automorphisms as in (*)). Now, a homomorphism ρ is called *rigid* if every deformation of ρ is trivial.

2.1.A. The space of homomorphisms. Denote by $\mathcal{R}$ the space of homomorphisms $\rho : \Gamma \to G$ and then take the quotient space $\bar{\mathcal{R}} = \mathcal{R}/\mathrm{Aut}\, G$. Then the rigidity of ρ can be expressed by saying that $\bar{\rho} \in \bar{\mathcal{R}}$ (i.e. the image of ρ under the tautological map $\mathcal{R} \to \bar{\mathcal{R}}$) is an *isolated* point of $\bar{\mathcal{R}}$. To be rigorous here one should be careful with the topology in $\bar{\mathcal{R}}$. The subtlety comes from possible non-compactness of the group $\mathrm{Aut}\, G$ which can make the space $\bar{\mathcal{R}}$ non-Hausdorff. Yet the difficulty is not serious and one can equate with little foundational work the following three properties.

1. ρ is rigid.

2. $\bar{\rho} \in \bar{\mathcal{R}}$ is an isolated point.

3. The dimension of $\bar{\mathcal{R}}$ at $\bar{\rho}$ equals zero. (This definition suggests $\dim_{\bar{\rho}} \bar{\mathcal{R}}$ for a measure of non-rigidity of $\bar{\rho}$ in the case this dimension is positive).

2.2. Rigidity of subgroups. A subgroup $\Gamma \subset G$ is called *rigid* if the inclusion $\Gamma \subset G$ is a rigid homomorphism in the above sense.

2.2.A. Strong rigidity of Mostow. A lattice Γ in G is called (strongly) *Mostow rigid* if for another lattice in an arbitrary (connected as usual) Lie group, say $\Gamma' \subset G'$, every isomorphism between the lattices, $\Gamma \leftrightarrow \Gamma'$ extends to a unique isomorphism of the ambient Lie groups, $G \leftrightarrow G'$.

An essentially equivalent rigidity property for locally homogeneous spaces V of finite volume says that every V' *with finite volume* which is homotopy equivalent to V is isometric to V.

2.2.A'. If $\Gamma \subset G$ is a *cocompact* lattice then the Mostow rigidity is stronger than the ordinary (local) rigidity. In fact, a small deformation $\Gamma_\varepsilon \subset G$ of Γ is again discrete and cocompact by a simple argument (an exercise to the reader). Then the strong rigidity provides automorphisms a_ε of G_ε sending $\Gamma = \Gamma_0$ to Γ_ε which are (by another simple argument) continuous in ε. But, amazingly, there exist non-cocompact Mostow rigid lattices which are not locally rigid (see [Th], $[Gr]_1$).

2.3. Rigidity of G-actions. Let a Lie group G smoothly act on a manifold W and let $A : G \times W \to W$ denote this action. We say that A is *rigid* if for every smooth one-parameter family of actions A_t of G on W there exists a one-parameter family of diffeomorphisms $h_t : W \to W$ which conjugate A to A_t for all t in a small interval $(-\varepsilon,\varepsilon) \subset \mathbb{R} \to t$.

2.3.A'. Example. Let Γ be a cocompact lattice in G and consider the (compact) G-homogeneous space $W = G/\Gamma$. The topology of this space does not change if we slightly deform Γ (this must be obvious to the readers who solved the exercise indicated in 2.2.A'). Thus deformations Γ_ε of Γ naturally correspond to deformations A_ε of the original action A of G on $W = G/\Gamma$. Next, one can easily see that a diffeomorphism h_ε of W changing the action A to A_ε is essentially the same thing as a conjugation of Γ to Γ_ε in G. Thus the rigidity of the action implies the rigidity of Γ with the additional property of the implied automorphism $a_\varepsilon : G \to G$ being *inner*. (Making a detailed proof amounts to unraveling the pertinent definitions concerning groups, homogeneous spaces etc.).

2.4. Rigidity of finite groups. If Γ is a *finite* group, then every homomorphism $\Gamma \to G$ is locally rigid. This is a classical fact and it can be seen from various angles. Here are some of them.

2.4.A. If G is a subgroup in GL_n, then homomorphisms of Γ to G induce those to GL_n that are *linear representations* of Γ. One knows since the last century that n-dimensional representations of finite (or compact) group form a finite set modulo conjugation in GL_n. This implies the rigidity in GL_n and with some extra work one can take care of all $G \subset GL_n$.

2.4.B. Consider two homomorphisms of Γ to G, denoted $\gamma \mapsto \bar{\gamma}$ and $\gamma \mapsto \hat{\gamma}$, and let us try to find a conjugation of $\bar{\Gamma}$ to $\hat{\Gamma}$. That is, we are after an element $g \in G$ satisfying $g\bar{\gamma}g^{-1} = \hat{\gamma}$, for all $\gamma \in \Gamma$, or, equivalently

$$g = \hat{\gamma} g \bar{\gamma}^{-1} \qquad (*)$$

The equation (*) should be thought of as the fixed point condition for the action

$$g \mapsto \hat{\gamma} g \bar{\gamma}^{-1} \qquad (**)$$

of Γ on G.

If $\bar{\Gamma} = \hat{\Gamma}$ then the obvious fixed point is $g = \mathrm{Id} \in G$. Now, if $\bar{\gamma}$ is close to $\hat{\gamma}$ for all $\gamma \in \Gamma$ we can invoke the following classical

2.4.B'. Stability theorem. *Let a finite (or compact) group Γ act on a smooth manifold W and $w_0 \in W$ be a fixed point of this action. Then w_0 is stable under small smooth perturbation of the action say A_ε of A, in the following sense. Every action A_ε close to $A = A_0$ has a fixed point $w_\varepsilon \in W$ which is continuous in ε for small ε.*

Idea of the proof. If the space W and the actions A_ε are linear, then the fixed point w_ε can be obtained by the averaging of the orbit $\gamma_\varepsilon(w_0)$ over Γ, where γ_ε denotes the action of Γ according to A_ε. (This is what is used in the representation theory mentioned above). Then the general nonlinear case follows by using a linearization of the action and a suitable implicit function theorem. Alternatively, one may use an invariant Riemannian metric on W (which can also be obtained by averaging an arbitrary metric) and then take the center of mass of the orbit $\gamma_\varepsilon(w_0)$ for such a metric (see [Kar] for a center of mass construction).

2.4.C. *If G is a compact group then every smooth action of G on a compact manifold W is rigid.* In fact, the diffeomorphisms h moving one action to another, say $\bar{g}$ to $\hat{g}$, are the fixed points of the action $h \mapsto \hat{g} \circ h \circ \bar{g}^{-1}$ of G on the space of maps $h : W \to W$. The existence of a fixed point in this (infinite dimensional) space can be obtained by the argument indicated in 1.4.B'., which implies the desired rigidity according to Mostow, Palais etc. (see [Mos]$_2$, [Pal]). Notice that this rigidity implies that for discrete (and hence finite) subgroups $\Gamma \subset G$.

2.4.D. On the failure of the strong rigidity. Two isomorphic finite subgroups Γ and Γ' in G do not have to be conjugate or be obtainable one from the other by an automorphism of G. A trivial example is that of two different representations of the cyclic group $\Gamma = \mathbb{Z}/p\mathbb{Z}$ into the unitary group $U(1) = S^1 \subset \mathbb{C}^\times$. Every such representation is determined by where a given generator $\gamma_0 \in \Gamma$ goes. The image of γ_0 is a p-th root of unity $\mathbb{C}^\times$ and by taking the representations $\gamma_0 \mapsto \alpha_0$ and $\gamma_0 \mapsto \alpha'_0$ for different roots of unity we obtain non-equivalent representations.

This example is not very convincing as the *images* of the representations are the same, provided the representations in question are injective (i.e. the roots α are primitive). Namely, the image consists of *all* p-th roots of unity in such a case. A more interesting situation arises for cyclic subgroups Γ in $U(2)$. Such a subgroup can be generated by a diagonal matrix $\begin{pmatrix} \alpha_0 & 0 \\ 0 & \beta_0 \end{pmatrix}$ where α_0 and β_0 are p-th roots of unity. Now, for a fixed α_0 we can vary β_0 thus obtaining non-conjugate subgroups in $U(2)$ (Two subgroups can be distinguished by the trace, $\gamma_0^i \mapsto \alpha_0^i + \beta_0^i$, $i = 1,2,\dots,p$, which is a function on $\Gamma = \{\gamma_0,\dots,\gamma_0^p\}$ invariant under conjugations in $U(2)$) .

One can see everything more geometrically by looking on how Γ acts on the unit sphere $S^3 \subset \mathbb{C}^2$. If this action is free (which is the case for most representations $\Gamma \to U(2)$) one takes the quotient manifold S^3/Γ which is called a *lens space* and which carries a (locally homogeneous) metric of constant curvature $+1$. The above discussion shows that two different lens spaces do not have to be isometric. (We suggest to the reader to work out a specific example and describe, geometrically, an invariant distinguishing S^3/Γ and S^3/Γ' for Γ and Γ' isomorphic to $\mathbb{Z}/5\mathbb{Z}$) .

2.4.D'. The geometry and topology of lens spaces goes much deeper than one might think. For example two such spaces can be homotopy equivalent but not diffeomorphic (see $[\mathrm{Mil}]_1$).

2.5. Rigidity of lattices in Abelian and nilpotent groups. Every isomorphism between two lattices in $\mathbb{R}^n$ is induced by a unique automorphism of $\mathbb{R}^n$. In fact, continuous automorphisms of $\mathbb{R}^n$ are just linear automorphisms. As every lattice $\Gamma \subset \mathbb{R}^n$ linearly spans $\mathbb{R}^n$, the homomorphisms $h : \Gamma \to \mathbb{R}^m$ uniquely extend to $\mathbb{R}^n$ by linearity : each $x = \sum_i c_i\gamma_i \in \mathbb{R}^n$ goes to $h(x) \underset{\text{def}}{=} \sum_i c_i h(\gamma_i) \in \mathbb{R}^m$. Of course, one should check that $h(x)$ does not depend on how x is combined out of γ_i , i.e. the equality $\sum_i c_i\gamma_i = \sum_i c'_i\gamma_i$ must imply $\sum_i c_i h(\gamma_i) = \sum_i c'_i h(\gamma_i)$. This is indeed true and easy to prove.

2.5.A. Let us indicate a more geometric way of extension of h from Γ to $\mathbb{R}^n$. First we consider the group $\Gamma_{\mathbb{Q}} \subset \mathbb{R}^n$ of Γ-*rational* points, where $x \in \mathbb{R}^n$ is called Γ-*rational* if $nx \in \Gamma$ for some $x \in \mathbb{Z}$. In other words

$$\Gamma_{\mathbb{Q}} = \bigcap_{n \in \mathbb{Z}} n^{-1}\Gamma ,$$

where

$$\alpha\Gamma \underset{\text{def}}{=} \{\alpha\gamma \mid \gamma \in \Gamma\} .$$

Now, every h extends from Γ to $\Gamma_{\mathbb{Q}}$ by $h(n^{-1}\gamma) = n^{-1}h(\gamma)$ for all $\gamma \in \Gamma$ and $n \in \mathbb{Z}$. Here again one must check that

$$n_1^{-1}\gamma_1 = n_2^{-1}\gamma_2 \Rightarrow n_1^{-1}h(\gamma_1) = n_2^{-1}h(\gamma_2) ,$$

but this is quite easy. Moreover, the extension

$$h_{\mathbb{Q}} : \Gamma_{\mathbb{Q}} \to \mathbb{R}^m$$

is a homomorphism which (by an extra easy argument) is continuous for the induced topology of $\Gamma_{\mathbb{Q}} \subset \mathbb{R}^n$. Thus $h_{\mathbb{Q}}$ extends by continuity to the desired homomorphism, say

$$h_{\mathbb{R}} : \mathbb{R}^n \to \mathbb{R}^m.$$

Exercise. Show that the existence of an extension of h from Γ to $\mathbb{R}^n$ remains valid for non-cocompact discrete subgroups Γ but then such an extension is non-unique.

2.5.B. The rigidity of Γ can be expressed in terms of the flat torus $T^n = \mathbb{R}^n/\Gamma$ as follows : every automorphism of $\Gamma = \pi_1(T^n)$ is induced by an affine automorphism of T^n that is a diffeomorphism preserving the Levi Civita connection of the flat metric on T^n. Furthermore, this automorphism is unique up to translations of T^n.

Remarks. (a) We can not replace the affine automorphisms by isometries. In fact as we deform a lattice $\Gamma \subset \mathbb{R}^n$ the isometry type of T^n may change.

(b) One can slightly modify and generalize the above as follows :

Every continuous map between flat tori, say $T^n \to T^m$ *is homotopic to an affine map and this affine map is unique up to translations of* T^n *on* T^m. (Here, affine maps don't have to be diffeomorphic, but just continuous sending geodesics to geodesics).

The proof of this mapping theorem is quite easy and we suggest the reader to work it out. A more difficult exercise to the reader is to prove a similar result for maps between arbitrarily compact flat Riemannnian manifolds.

We shall see later on in §4 that a similar mapping theorem holds true for some (non flat!) locally symmetric spaces of non-compact type but the required proof is significantly deeper.

2.5.C. Nilpotent groups. A group G is called nilpotent of (nilpotency) degree k if for arbitrary $k+1$ elements $g_0, g_1, \dots, g_k$ the iterated commutator of g_i equals id,

$$[g_0,[g_1,\dots,[g_{k-1}, g_k]]] = \text{id} .$$

Thus "nilpotent of degree one" means "abelian" as

$$[g_0,g_1] \underset{\text{def}}{=} g_0 g_1 g_0^{-1} g_1^{-1} = \text{id} ,$$

for all g_0 and g_1 in G. The nilpotency of degree 2 requires

$$[g_0,[g_1, g_2]] = \text{id}$$

for all g_0, g_1 and g_2 in G and so on. Finally, just "nilpotent" means "nilpotent of some degree k ".

2.5.C'. Example. The group Δ^{k+1} of upper triangular $(k+1) \times (k+1)$-matrices with unit diagonal entries is nilpotent of order k. This is seen immediately. The first interesting example is the Heisenberg (abc)-group

$\Delta^3 = \left\{ \begin{pmatrix} 1 & a & c \\ 0 & 1 & b \\ 0 & 0 & 1 \end{pmatrix} \right\}$. Here the matrices $\underline{a} = \begin{pmatrix} 1 & a & 0 \\ 0 & 1 & 0 \\ 0 & 0 & 1 \end{pmatrix}$, $\underline{b} = \begin{pmatrix} 1 & 0 & 0 \\ 0 & 1 & b \\ 0 & 0 & 1 \end{pmatrix}$ and $\underline{c} = \begin{pmatrix} 1 & 0 & c \\ 0 & 1 & 0 \\ 0 & 0 & 1 \end{pmatrix}$ satisfy $[\underline{a},\underline{c}] = [\underline{b},\underline{c}] = 1$ and $[\underline{a},\underline{b}] = \begin{pmatrix} 1 & 0 & ab \\ 0 & 1 & 0 \\ 0 & 0 & 1 \end{pmatrix}$.

2.5.C_1. <u>Malcev uniqueness (rigidity) theorem</u>. *let* G *and* G' *be simply connected nilpotent Lie groups and* $\Gamma \subset G$ *and* $\Gamma' \subset G'$ *be lattices. Then every isomorphism* $\Gamma \leftrightarrow \Gamma'$ *uniquely extends to an isomorphism* $G \leftrightarrow G'$.

The corresponding existence theorem is also due to Malcev. It reads

2.5.C_2. *For every finitely generated torsionless nilpotent group* Γ *there exists a simply connected Lie group* G *containing* Γ *as a lattice.*

The idea of the proof of 2.5.C_1 is the same as the one used in the abelian case. First one defines $\Gamma_{\mathbb{Q}} \subset G$ as the set of all $g \in G$ satisfying $g^n \in \Gamma$ for some $n \in \mathbb{Z}$, and one shows that $\Gamma_{\mathbb{Q}}$ is a dense subgroup in G . Then one extends an isomorphism from Γ to $\Gamma_{\mathbb{Q}}$ by a purely algebraic argument. The final extension from $\Gamma_{\mathbb{Q}}$ to G is obtained by continuity. Notice that, as in the abelian case, everything works here for an arbitrary homomorphism $h : \Gamma \to G'$, and allows a unique extension to a homomorphism between the Lie groups, say $h_{\mathbb{R}} : G \to G'$.

The proof of the Malcev existence theorem is similar to how one builds real numbers starting from integers. First one abstractly constructs the group $\Gamma_{\mathbb{Q}}$ by introducing the "roots" $\gamma^{1/n}$, for all $\gamma \in \Gamma$ and $n \in \mathbb{Z}$. The group $\Gamma_{\mathbb{Q}}$ is then given an appropriate topology and $G = \Gamma_{\mathbb{R}} \supset \Gamma_{\mathbb{Q}}$ is obtained by completing $\Gamma_{\mathbb{Q}}$.

The actual proof of the Malcev theorem is somewhat involved and we do not explain it here (see [Mal]). But we want to indicate a somewhat different geometric construction of a Lie group starting from a discrete nilpotent group Γ . The rough idea is to view Γ as a discrete set looked at from a certain point v "outside" Γ . As the view point v goes further and further away from Γ the "visual gaps" between points in Γ become smaller and smaller. Then, in the limit for $v \to \infty$, we shall see no gaps at all but rather a continuous object, which, in fact, can be given a structure of a nilpotent Lie group. In more technical terms this limit group is obtained with some auxiliary (word) metric ρ on Γ by taking

$$\Gamma^* = \lim_{n \to \infty} (\Gamma, n^{-1}\rho) ,$$

where one uses the (pointed) *Hausdorff topology* to go to the limit. (See [G-L-P], $[Gr]_2$). The above limit group Γ^* is not the same as $G = \Gamma_{\mathbb{R}}$ obtained by completing $\Gamma_{\mathbb{Q}}$, though both are Lie groups of the same dimension and nilpotency degree. Yet, Γ^* is not always isomorphic to G (see $[Pan]_1$ for an extensive discussion).

Examples. (a) If Γ is a free abelian group of rank n, then, of course, $\Gamma^* = \mathbb{R}^n$. Yet even in this case there is no canonical embedding of Γ to Γ^*.

(b) Let $\Gamma \subset \Delta^{k+1}$ consist of the triangular matrices with integer entries. Then $\Gamma_{\mathbb{Q}}$ consists of the matrices with rational entries (this is easy to show) and the Malcev theorem becomes especially transparent. Also in this case Γ^* is isomorphic to $\Gamma_{\mathbb{R}} = \Delta^{k+1}$.

Remarks. (a) The above discussion suggests the existence of certain universal constructions (functors) leading from discrete to continuous groups, but no direct generalization of $\Gamma > \Gamma_{\mathbb{R}}$ and $\Gamma > \Gamma^*$ seems possible for non (virtually) nilpotent groups, such as lattices in semisimple groups. Yet there are other constructions available as we shall see in the course of these lectures.

(b) The geometry of lattices in simply connected solvable Lie groups is similar to that for nilpotent groups although the final answers are less functorial. An interested reader may consult $[\text{Mos}]_3$, [Rag].

2.6. Representations of finitely presented groups. Let Γ be given by a finite presentation $\{\gamma_1,\dots,\gamma_k \mid r_1,\dots,r_\ell\}$, where $\gamma_1,\dots,\gamma_k$ are generators of Γ and $r_1,\dots,r_\ell$ are relations. Recall that every relation has the form

$$\gamma_{i_1}^{\varepsilon_1} \gamma_{i_2}^{\varepsilon_2} \cdots \gamma_{i_n}^{\varepsilon_n} = 1,$$

where ε_i are arbitrary integers, where the values of the indices i_j run through $1,\dots k$ possibly with repetitions (so that n can be any number) and where 1 denotes the neutral element. Then every homomorphism $\Gamma \to GL_n$, denoted $\gamma \mapsto \bar{\gamma}$, is determined by k matrices $\bar{\gamma}_1,\dots\bar{\gamma}_k$ in GL_n which satisfy the relations r_i with the matrices $\bar{\gamma}_i$ in place of γ_i. Notice that a single matrix relation amounts to n^2 algebraic equations imposed on the entries of the matrices $\bar{\gamma}_i$. As every $\bar{\gamma}_i$ has n^2 entries, the set of all representations $\Gamma \to GL_n$ appears as an algebraic subvariety $\mathcal{R}$ in the kn^2-dimensional Euclidean space given by ℓn^2 equation. Thus the expected dimension of $\mathcal{R}$ is $(k-\ell)n^2$. Recall that the real object of interest is not $\mathcal{R}$ but $\mathcal{M} = \mathcal{R}/GL_n$ where GL_n acts on $\mathcal{R}$ by conjugation. For $k \geq 2$ the action is free almost everywhere on the kn^2-dimensional space containing $\mathcal{R}$. Thus the expected dimension of $\mathcal{M}$ is $(k-\ell-1)n^2$.

If we work over *complex* numbers and our GL_n is, in fact, $GL_n\mathbb{C}$ then this heuristic computation does give rigorous results. Namely, if $k-\ell \geq 2$, then the dimension of $\mathcal{M}$ is at least n^2 at every point (by elementary intersection theory, see [Sha]). In particular, *no complex representation of* Γ *is rigid.*

Notice that the above argument remains valid for all complex Lie groups G in place of $GL_n\mathbb{C}$ and the conclusion often holds true for real groups as well. Thus we see that the rigidity of $\Gamma \subset G$ implies that the number ℓ of relations of Γ is almost as large as the number k of generators of Γ, namely

$$\ell \geq k-1.$$

2.6.A. Example. $G = SL_2\mathbb{R}$. This group has dimension 3 and so the expected dimension of $\mathfrak{M}$ is $3(k-\ell-1)$. In particular, if Γ is the fundamental group of a surface of genus g , then $k = 2g$ and $\ell = 1$, and the expected dimension of $\mathfrak{M}$ is $6g-6$. Now, a *discrete cocompact* surface group Γ in $SL_2\mathbb{R}$ gives us a surface V of constant negative curvature, that is

$$V = SO(2) \setminus SL_2\,\mathbb{R}/\Gamma .$$

The space of such surfaces of genus $g \geq 2$ is known (by elementary Teichmüller theory) to have dimension 6g-6 . Thus our heuristic computation gave us the right answer for $g \geq 2$. Notice that this answer implies non-rigidity of (most) lattices in $SL_2\mathbb{R}$.

2.6.B. Example : $G = SL_2\mathbb{C}$. For this group the corresponding symmetric space $X = SU(2) \setminus SL_2\mathbb{C}$ is the 3-dimensional hyperbolic space. Thus lattices Γ in $SL_2\mathbb{C}$ correspond to 3-dimensional manifolds (*orbifolds*, if Γ has torsion) V of constant negative curvature and finite volume. If V is compact then the corresponding Γ (cocompact in this case) admits a presentation with the number ℓ of relations equal to k, the number of generators. In fact, this is true for the fundamental group Γ of an arbitrary closed 3-manifold. To see this take a Morse function on V with a single maximum and single minimum. Then the number k of critical points of index 1 equals the number ℓ of the points of index 2, since

$$1 + k - \ell - 1 = \chi(V) = 0 .$$

It follows that the associated Morse complex has k one-cells and ℓ two-cells, which gives us the desired presentation of $\Gamma = \pi_1(V)$. Now, our heuristic computation gives

$$\dim_{\mathbb{C}} \mathfrak{M} = 3(k-\ell-1) = -3$$

which suggests that $\mathfrak{M}$ does not exist at all (or reduces to a single point corresponding to the trivial representation of Γ).

To see what really happens we consider a non-cocompact lattice Γ without torsion and look at the corresponding non-compact manifold V . One knows (see [Th], $[Gr]_1$) that this V is diffeomorphic to a compact manifold, say $\bar{V}$, whose boundary is the union of several, say m , 2-tori. The Euler characteristic of this V is (by an obvious argument)

$$\chi(V) = -m ,$$

and the above discussion gives a presentation of $\Gamma = \pi_1(V)$ with k generators and ℓ relation, such that $k = \ell + m$. (By attaching m solid tori to the boundary one adds m relations to Γ . But the resulting manifold V' is compact and so its fundamental group Γ' with $k' = k$ and $\ell' = \ell + m$ has $k' = \ell'$. This gives an explanation to the relation $k = \ell + m$) . Now our heuristic argument predicts that

$$\dim_{\mathbb{C}} \mathfrak{M} = m - 3 ,$$

but the correct answer is

$$\dim_{\mathbb{C}} \mathfrak{M} = m ,$$

where the dimension is measured at the point in $\mathfrak{M}$ corresponding to our Γ (see [Th]). In particular, cocompact lattices are rigid while non-cocompact ones are not. (However, all lattices in $SL_n\mathbb{C}$ are Mostow rigid, see §3).

<u>Remark.</u> The (rigidity) equality $\dim_{\mathbb{C}} \mathfrak{M} = 0$ for cocompact lattices in $SL_2\mathbb{C}$ should be thought of as a kind of uniqueness theorem (compare Malcev theorems above). The corresponding existence result (which is by far deeper than the rigidity and is due to Thurston) claims that a compact 3-manifold satisfying certain topological conditions admit metrics of constant negative curvature (see [Th]). Notice that the topological conditions needed for Thurston's existence proof do not seem the best possible. The best possible result (may be, too good to be true) would be as follows.

<u>Conjecture</u>. *A closed 3-manifold* V *admits a metric of constant negative curvature if and only if* $\pi_2(V) = \pi_3(V) = 0$ *and every solvable subgroup in* $\pi_1(V)$ *is cyclic.*

2.6.C. <u>Rigidity of lattices and the rigidity of polyhedra.</u> There is formal similarity between the deformation problem for discrete subgroups and that for polyhedra in the spaces X of constant curvature. For example, let P be a convex polyhedron in $\mathbb{R}^3$ and let us try to deform P without changing the geometry of its 2-faces. Of course, there are *trivial* deformations corresponding to the rigid motions of P (that come from the group $\operatorname{Iso} \mathbb{R}^3$), but we are after the real deformations which change the dihedral angles between the faces. In fact, every such deformation is given by the numbers attached to the edges of P and measuring (the change of) these angles. Denote by e the number of the edges and observe that every vertex of P imposes 3 relations on these angles. For example, if there are exactly 3 faces adjacent to some vertex of P, then all three dihedral angles are uniquely determined by the planar angles of the faces themselves at this point. If there are 4 faces, then there is one degree of freedom left and so on.

By the Euler characteristic formula the numbers of the vertices, faces and edges satisfy

$$v - e + f = 2 .$$

Furthermore, if all faces are triangular, then, obviously, $f = \frac{2}{3} e$ and so $e - 3v = -6$, which *suggests* there is no non-trivial deformations of P. In fact, the above computation "overcounts" the number of relations (as also happens for our heuristic count of the dimension of the deformation spaces in $SL_2\mathbb{C}$). To make the count more precise, we fix a 2-face F_0 of P and observe that the angles at the three edges of F_0 are uniquely determined by the remaining dihedral angles. Thus we reduce the situation to $k = e - 3$ parameters and $\ell = 3v - 9$ relations (as we do not need anymore the relations coming from the vertices of F_0). Now we have a perfect match

$$k = \ell ,$$

which suggests both the rigidity (uniqueness) as well as the existence of polyhedra with given geometry of the faces. In fact, the rigidity is the classical theorem of Cauchy and Dehn and the existence is also valid by the work of A.D. Alexandrov (see [Ale]).

2.6.C_1. Reflection groups in H^2. Let P be a convex n-gon, $n \geq 5$, in the hyperbolic plane H^2 all of whose angles are 90°. Then the k reflexions $\gamma_1,...,\gamma_k$ of H^2 with respect to the edges of P (or rather the straight lines ℓ_i extending the edges) generate a discrete group Γ with the fundamental domain P. (Notice that P is a distinguished fundamental domain as it is bounded by the lines ℓ_i = Fixed-point-set-of γ_i). Now we want to deform Γ by deforming P keeping all angles 90°. The geometry of such a P is determined by the length of the edges, that are n real numbers, which are subject to three relations. For example, every triangle P in H^2 is uniquely (up to Iso H^2) determined by its angles (of course these can not be all 90°). Thus the space $\mathfrak{M}$ for Γ has $\dim \mathfrak{M} = n-3$.

2.6.C_2. Reflection groups in H^3. Let us count the dimension of the space of deformations of a convex polyhedron $P \subset H^3$ which do not change the dihedral angles of P. We assume here that P is dual to a simplicial polyhedron, i.e. there are exactly 3 edges at every vertex of P. Then the deformation is controlled by the lengths of the edges, where the three edges at a fixed vertex V_0 are determined by the rest. Thus we have $k = e-3$ parameters restricted by $\ell = 3(f-3)$ relations, which gives us the identity $k = \ell$ (since $v-e+f = 2$ and $v = \frac{2}{3}e$). This identity *suggests* both the rigidity (uniqueness) or P as well as the existence theorem, and a theorem by Andreev (see [And], [Th]) confirms this prediction under some mild combinatorial restrictions on P.

Now let P have all dihedral angles 90°. Then the group Γ generated by the reflections around the faces of P is discrete and cocompact (we have assumed all along P is compact) and the above-mentioned Andreev theorem insures the rigidity of Γ. Moreover, the existence part of Andreev's theorem provides a group Γ whose fundamental domain is combinatorially isomorphic to a given abstract polyhedron P satisfying certain necessary conditions.

Finally notice that the structure of the equations in the Andreev (Thurston) theory is very much similar to that in the Cauchy-Dehn-Alexandrov theorem and the proofs in both cases follow the same strings of ideas.

2.7. Local rigidity theorem for (semi)simple Lie groups. *Let* Γ *be a lattice in a non-compact simple Lie group* G *which is not locally isomorphic to* $SL_2\mathbb{R}$ *or to* $SL_2\mathbb{C}$. *Then* Γ *is locally rigid. Furthermore, if* Γ *is ocompact, then the rigidity remains valid for* . $SL_2\mathbb{C}$ *and* $PSL_2\mathbb{C}$. *(These are the two groups locally isomorphic to* $SL_2\mathbb{C}$ *).*

We do not prove this rigidity in our lectures.

2.7.A. Historical remarks and references. The first rigidity theorem, due to Selberg [1960], applies to $\Gamma = SL_n\mathbb{Z} \subset SL_n\mathbb{R}$ for $n \geq 3$. His proof is geometric in nature and is based on the study of maximal Abelian subgroups (isomorphic to $\mathbb{Z}^{n-1}$) in Γ (see [Sel]). The rigidity of cocompact lattices in O(n,1) is due to Calabi [1961] and the cocompact Hermitian case (where the corresponding symmetric space is Hermitian) is covered by a theorem of Calabi and Vesentini [1960] (see [Cal], [Cal-Ves]). The general cocompact case is due to A. Weil [1962] (see [Wang] and the end of Ch. VII in [Rag]).

The method of Calabi-Vesentini and Weil is based on Hodge theory and Bochner type inequalities.

2.7.B. The above rigidity theorem extends to semi-simple Lie groups without compact normal subgroups where one should take special care of the components locally isomorphic to $SL_2\mathbb{R}$ and $SL_2\mathbb{C}$.

2.7.C. If G is a non-compact simple Lie group of dimension > 6 (i.e. bigger than $SL_2\mathbb{R}$ and $SL_2\mathbb{C}$) then the lattices Γ in G appear rigid many times over. In fact by looking at the list of known algebraic properties of such Γ one could naturally conclude that no such Γ may exist at all.

Now, if the mere existence of Γ is a miracle, the existence of a deformation would be a miracle squared. This makes the rigidity theorem look rather shallow, and one expects here deeper "overrigidity" results.

2.8. Mostow rigidity theorem. *Let* G *be a non-compact simple Lie group of dimension* > 3 *(i.e. not locally isomorphic to* $SL_2\mathbb{R}$ *). Then every lattice* $\Gamma \subset G$ *is Mostow rigid.*

We shall prove this theorem for some G in §§ 3 and 4.

2.8.A. Let us formulate the Mostow theorem in terms of locally symmetric spaces.

Let V *and* V' *be complete locally symmetric spaces of non-compact type (i.e. with Ricci* < 0 *) of finite volume, such that* Vol V = Vol V'. *If* V *is irreducible of dimension* > 3, *then every isomorphism between the fundamental groups,* $\alpha : \pi_1(V) \to \pi_1(V')$ *is induced by a unique isometry of* V *onto* V'.

2.8.B. Remarks and corollaries. (a) As we fix no reference points in V and V' the fundamental groups are not truly defined and α should be thought of as not as an actual isomorphism but as an isomorphism up to conjugations in $\pi_1(V)$ and $\pi_1(V')$.

(b) An immediate corollary to the Mostow theorem reads

The group of the exterior automorphisms of $\Gamma = \pi_1(V)$, *that is* Aut Γ/Inn Aut Γ, *is canonically isomorphic to the isometry group of* V.

To see this we apply the theorem to the case V = V'.

Next we observe that the isometry group of V is *finite*. In fact this finiteness is a general (and well known) property which holds true for all complete Riemannian manifolds V of finite volume and negative Ricci curvature (Bochner theorem, see [B-Y]). Finally we conclude that the group Ext Aut Γ is finite as well.

(c) The equality Vol V = Vol V' is a normalization condition. We could equally use another condition, say

$$\text{Scal Curv } V = \text{Scal Curv } V'$$

(d) The Mostow theorem extends with little trouble to the *semi*-simple groups G (which correspond to reducible locally symmetric spaces V) if one takes proper care of $SL_2\mathbb{R}$-components of G.

2.8.C. References. The first instance of the Mostow rigidity was proven in 1967 for compact spaces of constant negative curvature (see $[\mathrm{Mos}]_1$). The general compact case appears in $[\mathrm{Mos}]_4$ in 1973. The Mostow rigidity for certain non-compact spaces (of $\mathbb{Q}$-rank one) is due to Prasad [Pra] and the general superrigidity of Margulis (see $[\mathrm{Mar}]_1$, $[\mathrm{Mar}]_2$). The basic idea of Mostow consists, in the study of the so-called *ideal boundary* ∂X of a non-compact symmetric space X, but the study of this boundary is quite different for the cases of $\operatorname{rank} X = 1$ and $\operatorname{rank} X \geq 2$.

2.8.C'. Recall that $\operatorname{rank} X$ is the maximal dimension of a *flat* in X that is a totally geodesic subspace isometric to a Euclidean space. Also notice that a symmetric space has rank one if and only if its sectional curvature is (strictly) negative, $K(X) < 0$.

According to the Cartan classification there are three series of non-compact rank 1 symmetric spaces and one exceptional space. They are

I. The real hyperbolic space H^n or $H^n_{\mathbb{R}}$. This is the only complete simply connected Riemannian manifold of constant sectional curvature.

II. The complex hyperbolic space $H^{2n}_{\mathbb{C}}$. This has an invariant Kähler metric of constant holomorphic curvature. One can also think of $H^{2n}_{\mathbb{C}}$ as the unit ball in $\mathbb{C}^n$ with the *Bergman* metric.

III. The quaternionic hyperbolic space $H^{4n}_{\mathbb{H}}$.

IV. The hyperbolic Cayley plane $H^{16}_{\mathbb{C}a}$.

We refer to $[\mathrm{Mos}]_4$ and $[\mathrm{Pan}]_2$ for a study of these spaces. Here we only mention the following important characteristic property.

The isometry group of every of the above hyperbolic spaces H *is transitive on the pairs of points* (x,y) *with a fixed distance* $\operatorname{dist}(x,y) = d$. This is equivalent to saying that the isotropy group $K_x \subset \operatorname{Iso} H$ of each point $x \in H$ (that is the maximal compact subgroup in $\operatorname{Iso} H$) is transitive on the unit tangent sphere of $S_x(1) \subset T_x(H)$. For $H = H^n_{\mathbb{R}}$ this K_x is the orthogonal group $O(n)$ naturally acting on $\mathbb{R}^n = T_x(H)$; for $H^{2n}_{\mathbb{C}}$ this K is $U(n)$; for $H^{4n}_{\mathbb{H}}$ this is the (unitary) symplectic group $Sp(n)$ (of linear transformations of the quaternionic space $\mathbb{H}^n$) and for $H^{16}_{\mathbb{C}a}$ this is the famous group Spin 9 acting on the Cayley plane $\mathbb{C}a^2 = \mathbb{R}^{16}$.

2.9. Margulis' (super)rigidity. Consider a discrete subgroup $\Gamma \subset G$ and a homomorphism of Γ into another Lie group, say $h : \Gamma \to G'$. Margulis' property, says, roughly speaking, that h extends to a (continuous !) homomorphism $\bar{h} : G \to G'$ unless the homomorphism h was somewhat "exceptional".

Here is a precise statement for simple Lie groups G.

2.9.A. *Let* G *be a non-compact simple Lie group which is not locally isomorphic to* $O(n,1) = \mathrm{Iso}\, H^n_{\mathbb{R}}$ *or to* $U(n,1) = \mathrm{Iso}\, H^{2n}_{\mathbb{C}}$ *and let* Γ *be a lattice in* G. *Then every linear representation* h *of* Γ, *that is a homomorphism* $h : \Gamma \to GL(N)$ *extends to a continuous homomorphism* $\bar{h} : G \to GL(N)$ *unless the image* $h(\Gamma)$ *is precompact in* $GL(N)$.

This theorem for $\mathrm{rank}_{\mathbb{R}} G \geq 2$ (where the underlying symmetric space has rank ≥ 2) was proven by Margulis in 1974 (see $[\mathrm{Mar}]_1$, $[\mathrm{Mar}]_2$, [Zim], $[\mathrm{Für}]_2$) using random walks on Γ and G and related ideas of ergodic theory. In the remaining cases, $G = Sp(n,1) = \mathrm{Iso}\, H^{4n}_{\mathbb{C}}$ and $G = F_4 = \mathrm{Iso}\, H^{16}_{\mathbb{C}a}$, this theorem was recently proven by K. Corlette (see $[\mathrm{Cor}]_1$), who used the method of harmonic maps (see §4).

2.9.B. Let us indicate a special case of superrididity in terms of locally symmetric spaces. Let V and V' be locally symmetric spaces with fundamental groups Γ and Γ' and let $h : \Gamma \to \Gamma'$ be a homomorphism. We want to find a *geodesic* map $\hat{h} : V \to V'$ which induces h on Γ (where h is taken up to conjugation as in 2.8.B.). Recall, that a map $V \to V'$ between Riemannian manifolds is called geodesic if the graph of this map is a totally geodesic submanifold in $V \times V'$. For example, geodesic maps $\mathbb{R}^m \to \mathbb{R}^n$ are the same as affine maps.

Now we suppose $V = K\backslash G/\Gamma$ where G satisfies the assumptions of the Margulis-Corlette theorem and let us explain how to construct $\hat{h}$ for an arbitrary locally symmetric V' of negative Ricci curvature. We use the adjoint linear representation of the group $G' = \mathrm{Iso}\, X'$ for the universal covering X' of V' and by composing this representation, say $G' \to GL(N)$ with $h : \Gamma \to \Gamma'$ and $\Gamma' \subset G = \mathrm{Iso}\, X'$ we obtain a linear representation $\Gamma \to GL(N)$ to which the super-rigidity applies and delivers a homomorphism $\bar{h} : G \to GL(N)$. Then it is not hard to show that $\bar{h}$ sends G to G' (i.e. to the image of G' in $GL(N)$ for the adjoint representation) and then we chose maximal compact subgroups $K \subset G$ and $K' \subset G'$ such that $\bar{h}(K) \subset K'$. Then we obtain a G-equivariant map $G/K \to G'/K'$ which moreover, can be made geodesic with an appropriate choice of $K' \supset \bar{h}(K)$ (see $[\mathrm{Mos}]_5$) and which descends to the required map $V \to V'$. (We suggest to the reader to fill in the details).

Notice, that the above argument can be reversed. Namely, every geodesic map $V \to V'$ lifts to the universal coverings and gives a map $X \to X'$ which induces a homomorphism $G \to G'$. For example, if X is geodesically embedded into X' then $\mathrm{Iso}\, X$ embeds into $\mathrm{Iso}\, X'$, as $\mathrm{Iso}\, X'$ is generated by geodesic symmetries of X around the points $x \in X' \subset X$.

In these lectures we do not present Margulis' proof but we explain how an arbitrary continuous map $\hat{h}_0 : V \to V'$ (inducing h on Γ) can be deformed to a harmonic map $\hat{h}$ and then we shall show following Corlette that this $\hat{h}$ is, in fact, totally geodesic for the case $G = Sp(n,1)$.

2.9.C. Let us explain the origin of the non-precompactness of the $h(\Gamma)$ condition in the super-rigidity theorem. First we observe that Γ admits "many homomorphisms" (see 2.9.C_1. below) onto finite groups F which embed into some $GL(N)$. For example, F acts by translations $f(\alpha) \mapsto f(\beta\alpha)$ on the space L of *all* functions $f : \Gamma \to \mathbb{R}$, where $N = \dim L = \mathrm{card}\, F$. It is perfectly clear that the resulting homomorphisms $h:\Gamma \to F = GL(N)$ do not usually extend to $G \supset \Gamma$. For instance, if G is simple, then every homomorphism $G \to GL(N)$ is either trivial or injective and so non-trivial non-injective homomorphisms h do not extend to G.

2.9.C_1. Let us explain where "many homomorphisms" $\Gamma \to F$ come from. The basic example is the group $\Gamma = SL_n\mathbb{Z} \subset SL_n\mathbb{R}$ which admits for every prime number p, a homomorphism $\Gamma \to F = SL_n\mathbb{F}_p$, where $\mathbb{F}_p$ denotes the finite field with p elements and $SL_n\mathbb{F}_p$ denotes the group of matrices with the entries from $\mathbb{F}_p$ and with determinant one. The homomorphisms $\Gamma \to F = SL_n\mathbb{F}_p$ is obtained by reducing the entries of integral matrices modulo p. Thus the kernel $\Gamma_p \subset \Gamma$ consists of the matrices having the diagonal entries equal $1 (\mathrm{mod}\ p)$ and outside the diagonal they are $0 (\mathrm{mod}\ p)$. (One can see directly that so defined Γ_p is a normal subgroup of finite index in Γ and then one constructs F as Γ/Γ_p without ever mentioning $\mathbb{F}_p$).

The above (mod p)-construction was extended by Selberg (see [Sel]) to all finitely generated subgroups in $GL(n)$. Thus Selberg has proven that Γ is *residually finite* that is *for every non-identity element* $\gamma \in \Gamma$ *there exists a homomorphism of* Γ *onto a finite group* F, *such that the image of* γ *in* F *is* $\neq \mathrm{id}$.

2.9.C_2. Let us indicate an example of a homomorphism $\Gamma \to GL(N)$ with an *infinite* precompact image. We start with the group $\Gamma_0 = \Gamma_0(q) \subset SL_n\mathbb{R}$ consisting of the matrices with the entries of the form $m+n\sqrt{q}$ where q is a fixed integer ≥ 2 and m and n run over $\mathbb{Z}$. Notice that Γ_0 is a dense countable subgroup in $SL_n\mathbb{R}$. Next we observe that the transformation $\alpha:\Gamma_0 \to \Gamma_0$ induced by

$$m+n\sqrt{q} \to m-n\sqrt{q} \qquad (*)$$

is an automorphism of Γ_0 as (*) is an automorphism of the ring

$$\mathbb{Z}(\sqrt{q}) = \{m+n\sqrt{q}\}_{m,n\in\mathbb{Z}}\,.$$

Notice that the composition of α with the original embedding $i:\ \Gamma_0 \subset SL_n\mathbb{R}$ defines another embedding, called $\bar{i} = \alpha \circ i : \Gamma_0 \to SL_n\mathbb{R}$. This new embedding is (highly!) discontinuous for the topology in Γ_0 induced by $i:\Gamma_0 \subset SL_n\mathbb{R}$ (this is obvious) and therefore $\bar{i}$ does not extend to $SL_n\mathbb{R} \overset{i}{\supset} \Gamma_0$.

Another interesting property here is that the embedding $i \oplus \bar{i} = \Gamma_0 \to SL_n\mathbb{R} \times SL_n\mathbb{R}$ has *discrete* image and this image has finite covolume in the group $SL_n\mathbb{R} \times SL_n\mathbb{R}$. (The discreteness of $i \oplus \bar{i} = \Gamma_0$ is rather obvious but the finite covolume property requires a little thought).

Now, we construct our example by taking the subgroup $\Gamma \subset \Gamma_0 \subset SL_n\mathbb{R}$ which consists of the linear transformations of $\mathbb{R}^n$ preserving the quadratic form

$$f = \sum_{i=1}^{n-1} x_i^2 - \sqrt{q}\, x_n^2\,.$$

We observe that

$$\Gamma = \Gamma_o \cap O(f) ,$$

where $O(f)$ is the subgroup in $SL_n\mathbb{R}$ consisting of *all real* matrices preserving f. Notice $O(f)$ is isomorphic to $SO(n,1)$ that is the group of transformations preserving

$$f_o = \sum_{i=1}^{n-1} x_i^2 - x_n^2 .$$

Then we observe that the second embedding $\bar{i}$ sends Γ onto the group

$$\Gamma' \underset{def}{=} \Gamma \cap O(\bar{f}) ,$$

where

$$\bar{f} = \sum_{i=1}^{n-1} x_i^2 + \sqrt{q}\, x_n^2 ,$$

and the pair $(i,\bar{i})$ *discretely* embeds Γ into $SO(f) \times SO(\bar{f})$. Now, we observe that $SO(\bar{f})$ is compact as it is (obviously) isomorphic to the usual special orthogonal group $SO(n)$ and so the discreteness of Γ in $SO(f) \times SO(\bar{f})$ implies the discreteness of Γ in $SO(f)$. (Projecting Γ from $SO(f) \times SO(\bar{f})$ to $SO(f)$ is just an instance of the step 3 in the construction of arithmetic groups indicated in 1.4.). It follows from the general theory of arithmetic groups that Γ is a lattice in $SO(f)$. But here, moreover,using the elementary fact that the only solution to the equation $f(x) = 0$ for $x \in \mathbb{Z}(\sqrt{q})$ is $x = 0$, one can easily show (without any general theory) that Γ is cocompact in $SO(f)$.

Thus we obtained a cocompact lattice $\Gamma \subset G = O(n-1,1)$ which admits an isomorphism on an (obviously dense) subgroup $\Gamma' \subset SO(n)$ and this isomorphism is by no means extendible to a homomorphism $SO(n-1,1) \to SO(n)$.

A judicious reader might notice that the group $O(n-1,1)$ is the one where the super-rigidity does not apply anyway. We leave to such a reader to work out a similar example for the group $O(n-2,2)$ which has rank 2 and to which the Margulis theorem applies.

2.10. Non-Archimedean superrigidity and arithmeticity. Margulis (see $[Mar]_1$, [Zim]) proved his super-rigidity theorem for homomorphisms $h : \Gamma \to G'$ where G' may be a p-*adic* Lie group. Namely he has shown in this case that the image $h(\Gamma) \subset G'$ is necessarily precompact. This result, together with the ordinary (Archimedean) super-rigidity was used by Margulis to deduce his famous

Arithmeticity theorem. *Every lattice in a simple Lie group* G *of* $\mathrm{rank}_{\mathbb{R}} \geq 2$ *is arithmetic.*

If $\mathrm{rank}_{\mathbb{R}}\, G = 1$ then there may be non-arithmetic lattices $\Gamma \subset G$. For example every group $O(n,1)$ contains such a lattice (see [G-P]) as well as $U(2,1)$ and $U(3.1)$ (see $[Mos]_6$). On the other hand, Corlette's super-rigidity proof with harmonic maps extends to the p-adic case (see [G-S]) which yields the arithmeticity theorem for $Sp(n,1)$ and F_4. The case of $U(n,1)$ for $n \geq 4$ remains open.

Finally, we notice that as on the previous occasions all of the super-rigidity and arithmeticity discussion extends to the *semi*-simple case if appropriate provisions are made for the "dangerous" O(n,1) and U(n,1)-parts of G.

2.11. On the proofs of the rigidity theorems. The methods to prove rigidity are often more interesting and have wider scope than a particular result they may yield. Here is a list of basic methods some of which are discussed with more details later on.

I. Asymptotic geometry of G and $\Gamma \subset G$. This is sometimes called *long range, large scale* or *large distance* geometry.

The idea is to look at Γ from infinitely far away and recapture G (or at least the essential of G) from what one can see of Γ with a properly adapted eye (see §3).

II. Random walk on G and on Γ. If Vol $G/\Gamma < \infty$ then the random walk on Γ, as the time parameter $\to \infty$, approximates the random walk on G. The idea of relating G and Γ via their random walks originates from the work of Fürstenberg (see $[\text{Für}]_1$, $[\text{Für}]_3$). The power of this idea was demonstrated by Margulis in the course of his proof of the super-rigidity theorem. An interested reader is referred to [Zim], $[\text{Mar}]_3$.

III. Representation theory. One can think of the representation theory of a group as a linearization of random walks. Here again the passage from G to Γ is best possible if Vol $G/\Gamma < \infty$ and one has an especially beautiful representation (action) of G on the space of L_2-functions on G/Γ. The most important link between G and Γ established up to-day by means of the representation theory is the equivalence of *Kazdan's* T-*property* for G and lattices in G. A quick (and hardly comprehensible) way to define T for G is by saying that the trivial representation of G is isolated (rigid) in the space of all unitary representations. Basic examples of groups having T are the simple Lie groups except O(n,1) and U(n,1). The reader can find this (and much more) in [Kaz], [Zim], [dlH-V].

IV. Geometry and combinatorics of subgroups in G of positive codimension. By intersecting a lattice $\Gamma \subset G$ with Lie subgroups $G' \subset G$ one can often produce "interesting" subgroups $\Gamma' \subset \Gamma$. Conversely, starting from certain subgroups $\Gamma' \subset \Gamma$ one can sometimes recapture the geometric pattern of the corresponding Lie subgroups and eventually, rebuild G out of Γ. This method goes back to the first paper by Selberg and then appears in Mostow's rigidity theorem for rank $r \geq 2$, where one uses Abelian subgroups of rank r (see $[\text{Mos}]_4$). These ideas are also important in Margulis' super-rigidity proof.

V. Subgroups of finite index in Γ. It may be sometimes easier (and still very useful) to construct a p-adic Lie group $G_p \supset \Gamma$ rather than the original (real) Lie group $G \supset \Gamma$. For example, the group $\Gamma = SL_n\mathbb{Z}$ naturally embeds into $G_p = SL_n\mathbb{Q}_p$ as the integers embed into the field $\mathbb{Q}_p$ of the p-adic numbers. Notice that the closure $\bar{\Gamma}$ of this Γ in G_p is an *open* subgroup (consisting of the matrices with integer p-adic entries). It follows that the Lie algebra of G_p can be recovered from Γ and then one can recover G_p itself. Now, $\bar{\Gamma}$ is a compact totally disconnected group and so it is obtained by completing Γ for the topology defined by some system of subgroups $\Gamma_i \subset \Gamma$, $i \in I$ of finite index. in fact, in our case $\Gamma = SL_n\mathbb{Z}$ one gets the right Γ_i by taking the integral matrices $\equiv 1 \pmod{p^i}$.

The difficulty arising in this method is a characterization of the subgroups Γ_i in purely group theoretic terms. This problem does not appear however if Γ satisfies the *congruence subgroup property* saying that every subgroup of finite index in Γ contains a congruence subgroup distinguished by the $\equiv 1$ (mod q)-condition. Then one has a very quick proof of the super-rigidity theorem, as was pointed out to the authors by Alex Lubotzki. A reader interested in congruence subgroups is referred to [Bas].

VI. Harmonic maps. These are best explained in the language of locally symmetric spaces. If, for example, V and V' are locally symmetric spaces of non-compact type then every homomorphism between the fundamental groups, $\pi_1(V) \to \pi_1(V')$, is induced by a continuous map, say $f_0 : V \to V'$, because the universal covering of V' is contractible. Next, if we assume V and V' are compact and we recall that the sectional curvature of V' is non-positive, then we are in a position to apply the Eells-Sampson theorem which provides a *harmonic* map $f : V \to V'$ homotopic to f_0 (see [E-S], $[\text{E-L}]_1$, $[\text{E-L}]_2$ and §4 in these lectures). Now, we want to prove that f is, in fact, a geodesic map, which is by far stronger condition than mere harmonic. This can be sometimes done by means of Bochner type formulas as explained in §4.

§ 3 - Asymptotic geometry of symmetric spaces and the Mostow rigidity theorem.

We explain in this section how to recapture the geometry of a (symmetric) space X in terms of algebraic properties of a discrete cocompact group Γ acting on X. Then we sketch the proof of the Mostow rigidity theorem for symmetric spaces X of rank 1 (i.e. of negative sectional curvature) and indicate a sharpening of this theorem for the spaces $X = H^{4n}_{\mathbb{H}}$ and $H^{16}_{\mathbb{C}a}$.

3.1. Let a discrete group Γ isometrically act on a complete Riemannian manifold X, such that the quotient space X/Γ is compact. Then every orbit $\Gamma(x_0) \subset X$, $x_0 \in X$, is a net in X, i.e. there exists $R_0 \geq 0$ such that every ball of radius R_0 contains a point of $\Gamma(x_0)$ (compare 1.2.). Furthermore, $\Gamma(x_0)$ is *uniformly discrete* in X, which means a uniform bound on the number of points of $\Gamma(x_0)$ in the ball: For every $x \in X$ and $R > 0$ the number of points of $\Gamma(x_0)$ in the ball $B_x(R) \subset X$ is bounded,

$$\#((\Gamma(X_0) \cap B_x(R)) \leq N = N(R).$$

To simplify the exposition we assume below that Γ acts freely on X. Then X can be identified with the universal covering of $V = X/\Gamma$ and $\Gamma(X_0)$ equals the pull-back of a point $v_0 \in V$ under the covering map $p : X \to V$.

The simplest property of X which can be expressed in terms of $\Gamma(x_0)$ (and eventually in terms of Γ itself) is the *asymptotic volume* of X. Namely, we fix a point $x \in X$ and look at the volume of the ball $B_x(R)$ as $R \to \infty$. We first observe that in our case the volume function

$$\mathrm{Vo}_x(R) \underset{\mathrm{def}}{=} \mathrm{Vol}\, B_x(R)$$

grows at most exponentially. In fact, since $X/\mathrm{Iso}\, X$ is compact, a simple argument shows that

$$\mathrm{Vo}_x(R + R') \leq C\,\mathrm{Vo}_x(R) \qquad (*)$$

for some constant $C = C(R')$ depending only on R' and the geometry of $X/\mathrm{Iso}\, X$. In particular,

$$\mathrm{Vo}_x(R) \leq (C(1))^{R+1}$$

(Notice that this bound on the volume is satisfied whenever Ricci $X \geq -\delta > -\infty$, see [G-L-P]).

Then we notice that the function $\mathrm{Vo}_x(R)$ changes by an at most bounded amount if we move the point x,

$$\mathrm{Vo}_x(R)/\mathrm{Vo}_{x'}(R) \leq \mathrm{const} < \infty.$$

In fact the above constant satisfies

$$\text{const} \leq C(d) \text{ for } d = \text{dist}(x, x'),$$

as

$$B_{x'}(R) \subset B_x(R + d).$$

Next we compare the volume function $Vo_x(R)$ with the number

$$Nu_x(\Lambda ; R) = \#(\Lambda \cap B_x(R))$$

for a given net $\Lambda \subset X$. Using the net properties and the fact that

$$\text{Vol } B_x(R) \leq \text{const } (R)$$

(which follows from the above (*)), one immediately sees that

$$Vo_x(R) \leq C\, Nu_x(\Lambda, R), \quad R > 0, \qquad (+)$$

for some positive constant $C = C(X, \Lambda)$. Next we observe that the volume of the unit ball $B_x(1) \subset X$ is bounded away from zero uniformly for all $x \in X$, since $X/\text{Iso}X$ is compact. Then we conclude that

$$Vo_x(R) \geq C\, Nu_x(\Lambda, R), \qquad (++)$$

for some C as above and for all $R \geq R_0$, where R_0 is another constant depending on X and Λ.

We express (+) and (++) together by the notation

$$V_0 \sim Nu,$$

which should convey the idea of similar asymptotic growths of the functions $Vo(R)$ and $Nu(R)$ for $R \to \infty$.

Now we are interested in the case $\Lambda = \Gamma(X_0)$ and we want to relate the function $Nu(\Gamma(X_0), R)$ to some characteristic of the group Γ itself.

The key to such a relation is the following definition.

3.1. A. Word metric. Let $\Delta = \{\gamma_1, \dots \gamma_k\}$ be a system of generators of Γ. Then for each $\gamma \in \Gamma$ we define $|\gamma|_\Delta$ as the length of the shortest word in $\gamma_1, \dots \gamma_k$ and $\gamma_1^{-1}, \dots \gamma_k^{-1}$ representing γ. Then we define a *metric* $\text{dist}_\Delta(\gamma, \gamma')$ on Γ by

$$\text{dist}_\Delta(\gamma, \gamma') = |\gamma^{-1} \gamma'|_\Delta.$$

It is easy to see that this is indeed a metric which is, moreover, left invariant on Γ,

$$\mathrm{dist}(\alpha\gamma, \alpha\gamma') = \mathrm{dist}(\gamma, \gamma')$$

In fact dist_Δ is the *maximal* left invariant metric on Γ, satisfying

$$\mathrm{dist}_\Delta(\gamma_i^{\pm 1}, \mathrm{id}) \le 1.$$

This dist_Δ is what is called the *word metric* associated to Δ.

3.1.A. Lemma. *Every two word metrics on Γ are bi-Lipschitz equivalent. That is for every two finite generating systems Δ and Δ' the ratio $\mathrm{dist}_\Delta/\mathrm{dist}_{\Delta'}$ is a bounded function on the set of distinct pairs of points γ and γ' in Γ and the ratio $\mathrm{dist}_{\Delta'}/\mathrm{dist}_\Delta$ is also bounded.*

Proof. As the system Δ' is finite the length function $|\gamma|_\Delta$ is (obviously) bounded on $\Delta' \cup (\Delta')^{-1}$. Then this bound (obviously) extends to all of Γ as Δ' generates Γ.

3.1.A' Remark. An equivalent way to express the lemma is to say that every *isomorphism* between finitely generated groups $f : \Gamma \to \Gamma'$ is bi-Lipschitz for arbitrary word metrics in Γ and Γ'. In fact, every *homomorphism* (obviously) is *Lipschitz*,

$\mathrm{dist}_{\Gamma'}(f(\alpha), f(\beta)) \le \mathrm{const}\; \mathrm{dist}_\Gamma(\alpha, \beta)$ for all α and β in Γ and some const ≥ 0.

Next, we need a similar lemma which relates the geometry of Γ and an orbit $\Gamma(x_0)$.

3.1.B. *Let Γ discretely freely and isometrically act on a complete Riemannian manifold X. If X/Γ is compact then the (bijective) map $\Gamma \to \Gamma(x_0)$ for $\gamma \to \gamma(x_0)$ is bi-Lipschitz. That is there exists $\varepsilon > 0$, such that*

$$\varepsilon \le \mathrm{dist}_\Delta(\gamma, \gamma')/\mathrm{dist}_X(\gamma(x_0), \gamma'(x_0)) \le \varepsilon^{-1},$$

where dist_Δ denotes some word metric on Γ, where dist_X denotes the metric on X and where (γ, γ') run over all pairs of distinct elements in Γ.

Proof. Let $\Delta = \{\gamma_1, \dots \gamma_k\}$ and let $\ell_1 \dots \ell_k$ be smooth geodesic loops in V with a given base point $v_0 \in V$ which represent the generators $\gamma_1, \dots \gamma_k$ of the group Γ identified with the fundamental group $\pi_1(V, v_0)$. Then we consider the pull-back $\tilde{\Gamma} \subset X$ of the union $\ell_1 \cup \dots \cup \ell_k \subset V$ under the covering map and observe that $\tilde{\Gamma}$ is a *connected graph* in X consisting of geodesic segments joining certain pairs of points in the orbit $\Gamma(x_0)$ for some point $x_0 \in X$ over v_0. Notice that these segments that are edges of the graph $\tilde{\Gamma}$ may accidentally meet at some points besides the vertices of $\tilde{\Gamma}$ which are the points $\{\gamma(x_0)\}$, $\gamma \in \Gamma$. This happens if some loop ℓ has a double point or if two loops intersect at a point $v \ne v_0$. If $\dim V \ge 3$, we can slightly perturbe ℓ_i in order to remove extra intersection. In general, we pretend these intersections are not there and treat $\tilde{\Gamma} \subset X$ as an abstract graph with the vertices $\{\gamma(x_0)\}, \gamma \in \Gamma$ and the edges corresponding our geodesic segments.

Then we introduce the *generalized word* metric $\widetilde{\mathrm{dist}}$ in $\Gamma = \Gamma(x_0) \subset \tilde{\Gamma}$ as the length of the shortest path between γ and γ' in $\tilde{\Gamma}$. Notice that this reduces to the ordinary dist_Δ if length $\ell_i = 1$ for $i = 1, \ldots, k$ and that $\widetilde{\mathrm{dist}}_\Delta$ is, obviously, Lipschitz equivalent to dist_Δ.

Now the proof of the lemma reduces to the following.

3.1.B'. Sublemma. *The metric* dist_X *is bi-Lipschitz equivalent to* $\widetilde{\mathrm{dist}}$ on $\tilde{\Gamma} \subset X$.

Proof. The inequality

$$\mathrm{dist}_X \leq \widetilde{\mathrm{dist}}$$

is obvious. In fact, $\mathrm{dist}_X(x, y)$ equals the length of the shortest path in X between the points x and y chosen in $\tilde{\Gamma} \subset X$, while the metric $\widetilde{\mathrm{dist}}$ requires such a path to lie in $\tilde{\Gamma}$ which makes it, a priori, longer.

To prove the reverse inequality we start with the shortest path $[x, y]$ in V between x, y, which is a geodesic segment of length $d = \mathrm{dist}_X(x, y)$. Then we subdivide $[x, y]$ by some points $x = t_0, \ldots t_m, t_{m+1} = y$ into the subsegments

$$[x, t_1], [t_1\, t_2], \ldots [t_m, y],$$

such that the length of each of these subsegments does not exceed 1 and the number of them (i.e. $m + 1$) is no more than d.

Finally, for each point t_i, $i = 1, \ldots, m$ we take the nearest point $\tilde{t}_i \subset \tilde{\Gamma}$ and observe that

$$\mathrm{dist}_X (t_i, \tilde{t}_i) \leq C,$$

where

$$C = \sup_{x \in X} \mathrm{dist}(X, \Gamma(x_0)) < \infty,$$

as $\Gamma(x_0)$ is a net in X. It follows, by the triangle inequality that

$$\mathrm{dist}_X (\tilde{t}_i, \tilde{t}_{i+1}) \leq 2C + 1$$

for all t_i, $i = 0, \ldots, m + 1$.

Now, we use the cocompactness of Γ on $\tilde{\Gamma}$ and conclude to the existence of a constant $\tilde{C}$, such that

$$\widetilde{\mathrm{dist}}(\tilde{t}_i, \tilde{t}_{i+1}) \leq \tilde{C} \quad \text{for } i = 0, \ldots, m + 1$$

It follows, that

$$\widetilde{dist}(x, y) \leq \tilde{C}(dist_X(x, y) + 1),$$

which yields the required inequality

$$\widetilde{dist} \leq \tilde{C}' \, dist_X \qquad \text{on } \Gamma(x_0)$$

with a somewhat larger constant $\tilde{C}'$ as both distances are $> \varepsilon \geq 0$ on distinct points of $\Gamma(x_0)$.

3.1.C. Let us summarize what we have achieved.

I. The volume growth of X is equivalent to the growth of $\Gamma(x_0) \subset X$ for the metric $dist_X$,

$$Vo(X ; R) \sim Nu(\Gamma(x_0) ; R).$$

II. The orbit $\Gamma(x_0)$ with the metric $dist_X$ is bi-Lipschitz to Γ with some word metric $dist_\Delta$.

Now using $dist_\Delta$ one can introduce the growth function $Nu(\Gamma, \Delta ; R)$ that is the number of points in the $dist_\Delta$-ball in Γ of radius R around $id \in \Gamma$.

To say it differently, $Nu(\Gamma, \Delta ;R)$ is the number of elements in Γ representable by Δ-words of length $\leq R$.

Unfortunately, the bi-Lipschitz equivalence of metrics does not lead to equivalent growth functions. In fact, the distance R becomes about CR under a bi-Lipschitz equivalence but $N(R)$ is not equivalent to $N(CR)$ for most functions $N(R)$. For example $\exp 2R$ is not equivalent to $\exp R$ in our sense. However, our growth functions in question have at most exponential growth. For example, the number of words of length $\leq R + 1$ does not exceed $2k$ (the number of words of length $\leq R$), where k is the number of generators. Thus

$$Nu(R + 1) \leq 2k \, Nu(R)$$

which is the required growth bound.

Using this one immediately sees that

$$\log Nu(CR) \leq \text{const} \log Nu(R)$$

that is

$$\log Nu(CR) \sim \log Nu(R).$$

Also one obtains this equivalence for the bi-Lipschitz equivalent metrics and then concludes to the following.

3.1.D. Theorem. *The volume growth function* $Vo(R)$ *of* X *and the growth function* $Nu(R)$ *for some word metric* $dist_\Delta$ *in* Γ *satisfy*

$$\text{Log Nu}(R) \sim \log \text{Vo}(R) \quad \text{for } R \to \infty,$$

that is

$$0 < \varepsilon \le \log \text{Nu}(R)/\log \text{Vo}(R) \le \varepsilon^{-1} < \infty$$

for all sufficiently large R. (we assume Γ is infinite and so Nu(R) and Vo(R) go to ∞ for $R \to \infty$)

3.1.D'. Remarks (a) This theorem was first stated and proven by A.Švarc in 1955, following a hint by Efremowitz (see [Efr], [Šv] Then this result was rediscovered by J. Milnor in 1968. (see [Mil]$_2$). In fact, the results of Švarc and Milnor are somewhat more precise than 3.1.D. For example, they show, that if the group Γ is free Abelian of rank k, then the balls $B_x(R) \subset X$ have

$$\text{Vol } B_x(R) \ge CR^k \text{ for } R \to \infty.$$

We suggest to the reader to furnish the proof which is a slight modification of the argument presented above. (Notice that

$$\log R^k \sim \log R^\ell$$

for all k, $\ell > 0$ and so 3.1.D is useless here).

(b) Theorem 3.1.D shows how, in principle, one can relate asymptotic invariants of X and Γ. But it seems infinitely far from recapturing X itself from Γ, rather than only the growth rate of Γ. Yet some ideas in the proof will serve in the Mostow theorem later on where we take into account the shapes of X and Γ as well as of their mere sizes.

3.2. Quasi-isometries of metric spaces. Let us formulate explicitely the basic relation which connects a Riemannian manifold X and a cocompact lattice $\Gamma \subset \text{Iso } X$.

3.2.A. Definition . Two metric spaces X and Y are called *quasi-isometric* if there exist nets $X' \subset X$ and $Y' \subset Y$ and a bi-Lipschitz equivalence $f : X' \to Y'$, where X' is given the metric dist_X and Y' comes with dist_Y.

3.2.A'. Basic example. Let X and Γ be as above and Γ is given some word metric. Then X and Γ are quasi-isometric. In fact, if Γ acts freely at some point $x_0 \in X$ one can take $\Gamma(x_0)$ as a net in X and use all of Γ as a net in itself. Then the orbit map $\gamma \mapsto \gamma(x_0)$ is bi-Lipschitz. This has been shown in the previous section for free actions and the general case requires a minor adjustment of the argument.

Remark. It is true, but not automatic, that the quasi-isometry is an equivalence relation. However, it is clear in the above situation, that if the same group Γ cocompactly and isometrically acts on two different manifolds X and Y then X and Y are quasi-isometric. For example, if two *compact* manifolds V. and

W have isomorphic fundamental groups, then the universal covering $\tilde{V}$ and $\tilde{W}$ are quasi-isometric. In particular, such $\tilde{V}$ and $\tilde{W}$ have similar volume growth,

$$\log \mathrm{Vo}(\tilde{V}\,; R) \sim \log \mathrm{Vo}(\tilde{W}\,; R)$$

3.2.B. Let us give a more flexible definition of quasi-isometry. For this we need multi-valued maps $f : X \dashrightarrow Y$, where $f(x)$ is a subset in Y for each $x \in X$. Such a map can be represented by its graph

$$\Gamma r_f = \{x, y \mid y \in f(x)\} \subset X \times Y$$

and its convenient to allow arbitrary subsets in $X \times Y$ corresponding to *partially* defined multivalued maps $X \dashrightarrow Y$, which are called *correspondences.* . We define the (Hausdorff) distance between correspondences f, and g, denoted $|f - g|$, as the Hausdorff distance between their graphs Γr_f and Γr_g in $X \times Y$, where $X \times Y$ is given the sup-product metric

$$\mathrm{dist}((x_1, y_1), (x_2, y_2)) = \max(d^X, d^Y)$$

for $d^X = \mathrm{dist}(x_1, x_2)$ and $d^Y = \mathrm{dist}(y_1, y_2)$. Also recall that the Hausdorff distance between subsets A and B in a metric space is the infimum of those ε for which A is contained in the ε-neighbourhood of B and, conversely, B is contained in the ε-neighbourhood of A.

An important equivalence relation between maps is $|f - g| < \infty$.

<u>Example</u>. Let $X' \subset X$ be a uniformly discrete net and let $v(x) \subset X$ consist of the nearest to x points in X' (generically f(x) consists of a single point but for some $x \in X$, f(x) contains more the one element). Then, clearly, this v, called the *normal projection* of X to $X' \subset X$, is equivalent to the identity map, that is $|\mathrm{id} - v| < \infty$.

The following definition of quasi-isometry is invariant under the above equivalence relation.

3.2.B'. A map (correspondence) $f : X \dashrightarrow Y$ is called *quasi-isometry* if it satisfies the following two conditions.

QI_1. There exist positive constants A and B such that every four points x_1 and x_2 in X and $y_1 \in f(x_1) \subset Y$ and $y_2 \in f(x_2) \subset Y$ satisfy the following relation

$$A^{-1}\mathrm{dist}(x_1, x_2) - B \le \mathrm{dist}(y_1, y_2) \le A\,\mathrm{dist}(x_1, x_2) + B, \qquad (*)$$

(where we assume x_1 and x_2 lie in the domain of the definition of f).

QI_2. The domain of definition of f is a net in X while the image of f is a net in Y. That is, the projections of $\Gamma r_f \subset X \times Y$ to X and Y are nets.

3.2.B". Example. Let $X' \subset X$ and $Y' \subset Y$ be uniformly discrete nets and $f' : X' \to Y'$ be a bijection. Then the related correspondence $f : X \dashrightarrow Y$ with $\Gamma_f = \Gamma_{f'} \subset X' \times Y' \subset X \times Y$ is a quasi-isometry if and only if f' is a bi-Lipschitz map.

There is the following converse to this statement. For an arbitrary quasi-isometry $f : X \to Y$ there exist nets $X' \subset X$ and $Y' \subset Y$ and a bi-Lipschitz map $f' : X' \to Y'$ equivalent to f. The proof is a trivial exercise on the notions of a net and a quasi-isometry.

With all these definitions we can modify our basic problem of relating the geometries of X and Γ as follows : which geometric properties of X (and Γ) are quasi-isometry invariant ? So far we have only one such invariant namely the (logarithm of the) volume growth of X. But we want much more, at least if X is a symmetric space. Namely we want to reconstruct all of the geometry of X in quasi-isometric terms. We shall see below that this is possible for some symmetric spaces X.

3.3. The group of quasi-isometries $\overline{\text{QIs}}(X)$. Composition of two quasi-isometries is not, in general, a quasi-isometry. In fact, if the image of $f : X \dashrightarrow Y$ misses the domain of definition of $g : Y \dashrightarrow Z$, then $g \circ f : X \dashrightarrow Z$ has empty graph in $X \times Z$ and so is *not* quasi-isometric. To remedy this situation we consider only *full quasi-isometries* $f : X \dashrightarrow Y$, where the domain of definition equals X and the image is Y. Now, there is no problem to compose these.

Composition of full quasi-isometries (obviously) is a full quasi-isometry.

Since the inverse of a full q.i. is again a full q.i. we have a group, denoted FQIs(X) for every metric space X. This group appears too big to be useful but there is a reasonable factor group, denoted $\overline{\text{QIs}}(X)$, where the relevant normal subgroup consists of the quasi-isometries equivalent to the identity map. In other words, one goes from FQIs to $\overline{\text{QIs}}$ by using the equivalence relation $|f - g| < \infty$ on FQIs.

3.3.A. There is an obvious homomorphism of Iso X to $\overline{\text{QIs}}(X)$. For example, if $X = \mathbb{R}^n$ then the kernel of this homomorphism consists of the parallel translations of $\mathbb{R}^n$ and so the image is isomorphic to O(n). It means, in plain words that an isometry f of $\mathbb{R}^n$ has $|f - id| < \infty$ if and only if f is a parallel translation. To see this, one should notice that the metric $|f - g|$ on isometries is equivalent to the ordinary

$$\sup_{x \in X} \operatorname{dist}(f(x), g(x)) .$$

3.3.A' Our next example is that of a symmetric space X with Ricci $X < 0$ (i.e. X has no compact and flat de Rham factors). Then the homomorphism

$$\text{Iso } X \to \overline{\text{QIs}}(X)$$

is injective. In fact, let X be an arbitrary complete simply connected Riemannian manifold with non-positive sectional curvature and let $f : X \to X$ be an isometry, such that $f \neq id$ and

$$\sup_{x \in X} \operatorname{dist}(x, f(x) = s < \infty.$$

Then (see [B-C-S]) X isometrically splits.

$$X = X' \times \mathbb{R}$$

such that the isometry becomes

$$f : (X', t) \mapsto (x', t + s).$$

3.3.A". Remark. Let us interpret the above result in terms of the group $G = \mathrm{Iso}\, X$. Namely, we observe that the condition $|f - id| < \infty$ for a left invariant metric in G is equivalent to

$$\mathrm{dist}(fg, g) = |g^{-1} f^{-1} g| \leq C < \infty$$

for all $g \in G$. Intuitively, this says that f "almost commutes" with all $g \in G$. For example if G is a discrete group, then this condition implies that the centralizer of f has finite index in G. Furthermore if G is a non-compact simple group, then f must be in the center of G. We leave the proofs of all this to the reader.

3.3.B. Let us evaluate the group $\overline{\mathrm{QIs}}$ for $X = \mathbb{R}$. We observe, that for every continuous bounded function $\varphi : \mathbb{R} \to \mathbb{R}$ such that $\inf |\varphi| > 0$, the integral $t \mapsto \int_0^t \varphi(\tau)\, d\tau$ is a quasi-isometry (in fact, a bi-Lipschitz homeomorphism) of $\mathbb{R}$. Two such quasi-isometries give the same element in $\overline{\mathrm{QIs}}$ if and only if the corresponding function φ_1 and φ_2 have

$$\sup_{t \in \mathbb{R}} \left| \int_0^t (\varphi_1 - \varphi_1)(\tau)\, d\tau \right| < \infty .$$

This shows that the group $\overline{\mathrm{QIs}}\, \mathbb{R}$ is infinite dimensional.

Exercice. Show that the above $\int \varphi$ represent all of the group $\overline{\mathrm{QIs}}\mathbb{R}$.

3.4. Rigid metric spaces. A metric space X is called QIs-*rigid* if the homomorphism

$$\mathrm{Iso}\, X \to \overline{\mathrm{QIs}} X$$

is bijective (Compare [Ul].

The notion of a rigid space sharpens Mostow rigidity. In fact, every isomorphism between discrete groups Γ_1 and Γ_2 is a quasi-isometry and if Γ_1 and Γ_2 are cocompact lattices in Iso X, then the quasi-isometry $\Gamma_1 \to \Gamma_2$ defines a quasi-isometry $X \to X$. If X is rigid, this is equivalent to an isometry, as required by Mostow's rigidity. (Our definition in 2.2.A. allows different spaces X and Y and one needs an additional theorem saying that two symmetric spaces of non-compact type are isometric if and only if they are quasi-isometric. Such a theorem can be proven but we do not concern ourselves here with this aspect of rigidity).

In view of 3.3.B. the rigidity of X may appear an extremely unlikely property. In fact, one sees with 3.3.B. that the Euclidean space $\mathbb{R}^n$, $n \geq 1$, is not rigid. We shall also see in 3.11.C that the *hyperbolic spaces* $H^n_{\mathbb{R}}$ *and* $H^{2n}_{\mathbb{C}}$ *are not QIs-rigid.* However, it seems that *apart from* $H^n_{\mathbb{R}}$ *and* $H^{2n}_{\mathbb{C}}$ *all noncompact irreducible symmetric spaces are rigid.*

This was conjectured for the spaces of rank ≥ 2 by Margulis about 15 years ago (according to a private communication by G. Prasad) but the proof of Margulis'conjecture is still not available. On the other hand the QIs-rigidity for $H^{4n}_{\mathbb{H}}$ and $H^{16}_{\mathbb{C}a}$ has been proven in [Pan]2 and the idea of the proof is explained below in 3.11.

3.4.A. Remark on non-rigid spaces. Even if X is non-QIs-rigid one may try to define Iso X in quasi-isometric terms. One possibility is to use *uniformly quasi-isometric families* F of maps $f : X \to X$. This means there exists a constant A depending on F, such that each $f \in F$ satisfy the q.i. inequality (*) of 3.2.B with this A and some B depending on f. One can show that the subgroup

$$\mathrm{Iso}\, H^n_{\mathbb{R}} \subset \overline{\mathrm{QIs}}\, H^n_{\mathbb{R}}\ ,\ n \geq 2$$

is a *maximal subgroup of uniform* quasi-isometries. That is, for each $\varphi \in \overline{\mathrm{QIs}} - \mathrm{Iso}$ the subgroup generated by Iso and this φ contains quasi-isometries with arbitrary large (Lipschitz) constant A. Probably Iso $H^n_{\mathbb{R}}$ is a *unique* (up to conjugation in $\overline{\mathrm{QIs}}$) maximal uniformly quasi-isometric subgroup subject to some extra property (e.g. being connected or something similar), and the same should be true for $H^{2n}_{\mathbb{C}}$. However, this weakened QIs-rigidity does not seem to directly imply Mostow's rigidity.

3.5. The sphere at infinity. Recall that a (geodesic) *ray* in a complete Riemannian manifold X is an isometric copy of $\mathbb{R}_+$ in X. On can think of a ray as of a *subset* $R \subset X$ isometric to $\mathbb{R}_+$ or, sometimes, as an isometric map $\mathbb{R}_+ \to X$. By the definition of the Riemannian metric in X, the length of every segment $[a, b] \subset R \subset X$ equals $\mathrm{dist}_X(a, b)$ and so R is a *geodesic* ray.

Call two rays R_1 and R_2 in X *equivalent* if the Hausdorff distance between them is *bounded.* Then define the *ideal boundary* ∂X as the set of the equivalence classes of rays in X. The point $s \in \partial X$ represented by a ray R in X is called the (ideal) *end* of R.

3.5.A. Remarks and examples (a) There are many alternative definitions of an ideal boundary (see [Gr]2, [Gr]3) but they all turn out to be equivalent for symmetric spaces of rank one which are studied in this section. On the other hand, our definition may become quite ugly for some manifolds. Yet it serves well for *simply connected* manifolds with *non-positive* sectional curvature (e.g. for symmetric spaces of non-compact type).

(b) Let see what happens to $X = \mathbb{R}^n$. Here two rays are equivalent iff they are parallel and point in the same direction. Therefore, $\partial\mathbb{R}^n$ can be identified with the sphere S^{n-1} of the "directions" in $\mathbb{R}^n$ represented by the unit vectors.

(c) One may think of $X \cup \partial X$ as a *compactifiction* of X. In fact, let $x_i \in X$ be a divergent sequence of points such that dist $(x_0, x_i) \to \infty$ for $i \to \infty$. Then we consider minimizing geodesic segments $[x_0, x_i] \subset X$ between x_0 and x_i and take a subsequence $[x_0, x_j]$ for which the unit tangent vectors τ_j to $[x_0, x_j]$ at x_0 converge to some vector τ in the unit tangent sphere $S_{x_0}(X)$. It is easy to see that the one -sided infinite geodesic R in X issuing from x_0 in the direction of τ is a (geodesic) *ray* in X as the length of every segment $[x_0 = 0, r] \subset R = \mathbb{R}_+$ equals $\text{dist}_X(x_0, r)$. Thus one thinks of (the end of) R as an *ideal* (sub)-*limit* of the sequence x_i.

3.5.B. <u>The case</u> $K(X) \le 0$. Let the sectional curvature K(X) of X be non-positive and let us also assume X is simply connected as well as complete. Then by the Cartan-Hadamard theorem (see [B-G-S]) every one sided infinite geodesic is a (geodesic) ray in X. Thus if we take all geodesic rays in X issuing from a given point $x_0 \in X$ we obtain a map of the unit tangent sphere $S_{x_0}(X)$ into ∂X. Now we have the following basic theorem also going back to Cartan and Hadamard.

3.5.B_1. *The above map say* $G_{x_0} : S_{x_0} \to \partial X$ *is a bijection. Moreover, the composed map* $G_{x_0}^{-1} G_{y_1} : S_{y_1} \to S_{x_0}$ *is a homeomorphism for every pair of points* x_0 *and* y_1 *in* X.

<u>Sketch of the proof</u>. The essential property of X one uses is the following.

3.5.B_2. <u>Convexity of dist</u>. *The function* $(x, y) \mapsto \text{dist}(x, y)$ *is convex on* $X \times X$. *That is for every two geodesics* $g_1 : \mathbb{R} \to X$ *and* $g_2 : \mathbb{R} \to X$ *the function* $\text{dist}(g_1(t), g_2(\theta))$ *is convex on* $\mathbb{R}^2$. (See [B-G-S]).

Now one immediately sees that the map $S_{x_0} \to \partial X$ is *injective* as for any two distinct rays issuing from x_0, the distance $\text{dist}(r_1(t), r_2(t))$ goes to infinity for $t \to \infty$. In fact, every *convex* function d(t) with $d(0) = 0$ goes to infinity unless it is identically zero.

Next, to see that our map G_{x_0} is surjective we take an arbitrary ray $R_1 \subset X$ issuing from $r_1 \in X$ and look at the geodesic segments $[x_0, r_1] \subset X$ for all $r_1 \in R_1$ and a fixed x_0. Then we let $r_1 \to \infty$ and we need to show that the segments $[x_0, r_1]$ converge (or at least subconverge) to a ray R_0 issuing from x_0, such that $R_0 \sim R_1$, which means bounded distance

$$\text{dist}(r_0, r_1) \le \text{const} < \infty$$

for all $r_0 \in R_0$ and $r_1 \in R_1$ satisfying $\text{dist}(r_0, x_0) = \text{dist}(r_1, x_0)$ (This is obviously equivalent to the boundedness of Hausdorff distance).

Denote by $\tau(r_1) \in S_{x_0}(X)$ the tangent vector to $[x_0, r_1]$ at x_0 and look at the resulting curve $R_1 = \mathbb{R}_+ \to S_{x_0}(X)$ for $r_1 \mapsto \tau(r_1)$.

3.5.B$_3$. Lemma. The curve $\tau(r_1)$, $r_1 \in [0, \infty]$ has finite length in the sphere $S_{x_0}(X)$.

Proof. The radial projection of $X - \{x_0\}$ to $S_{x_0}(X)$ contracts every tangent vector $\tau_1 \in S_{x_1}(X)$ by the factor at least $(\text{dist}(x_0, x_1))^{-1}$ as easily follows from the convexity of the distance function between the segment $[x_0, x_1]$ and nearby segments x_0, x_1', for infinitesimal pertubations x_1' of x_1 in the direction of τ_1.

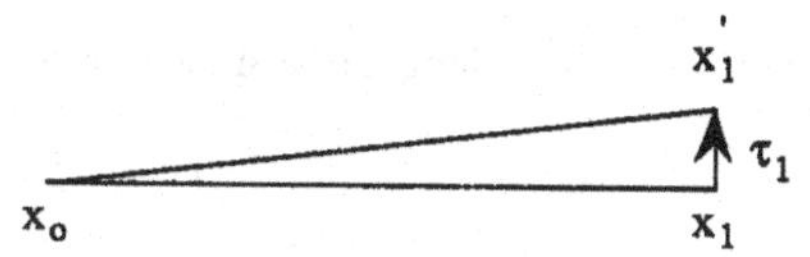

Moreover, this contraction is even stronger for all vectors τ_1 having small angle α_1 with the segment $[x_0, x_1]$. Namely it is of the order $\alpha_1(\text{dist}(x_0, x_1))^{-1}$.

For example, the vectors tangent to $[x_0, x_1]$ go to zero under the (differential of the) radial projection.

Now, the angle α_1 in our case can be again estimated using the convexity of the distance between the segments $[x_0, r_1]$ and $[y_1, r_1]$,

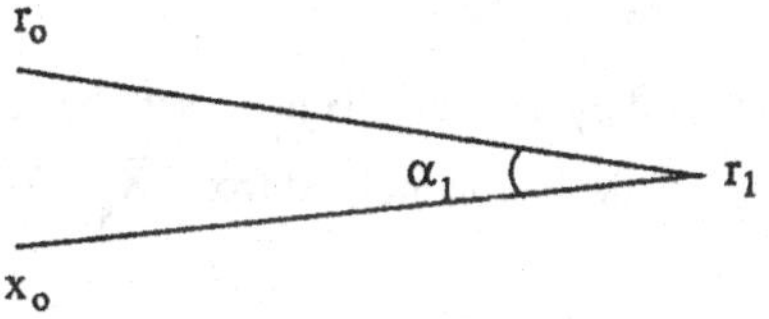

This convexity shows that

$$\alpha_1 < c(\text{dist}\,(x_0, r_1))^{-1}$$

and so the length of our curve is bounded by the integral

$$\int_0^\infty c\left(\text{dist}\,(x_0, r_1)\right)^{-2} dr_1 .$$

As $\text{dist}\,(x_0, r_1) \geq r_1 - \text{dist}\,(x_0, y_1)$, this integral converges for $r_1 \to \infty$ and the lemma follows.

Exercise. Show that the above curve in the sphere has length $< \pi$.

3.5.B_4. Now we see that $\tau(r_1)$ converges in $S_{x_0}(X)$ to some vector τ_0 and we take the ray R_0 defined by τ_0. Again, the convexity of the distances from $[x_0, r_1]$ to R_1 shows that the distance from R_0 to R_1 is bounded (by $\mathrm{dist}(x_0, y_1)$) and so $R_0 \sim R_1$ as required. The final *homeomorphism* property follows from the principle of L.E.J. Brower (applied to $G_{x_0}^{-1} G_{y_1}$ and $G_{y_1}^{-1} G_{x_0}$) claiming that every effectively defined map is continuous (The readers who do not trust this principle may check the continuity by looking carefully at the above (effective !) construction of R_0).

3.5.C. *Conformal boundary* $\partial_\varphi X$. Let us give another definition of the ideal boundary of X using a metric on X conformal to the original metric g on X. Such a conformal metric is $g_{new} = \varphi^2 g$ where φ is a continuous positive function on X. The new length of a smooth curve $x(t)$, $a \le t \le b$, in X is given by the integral

$$\mathrm{length}_{new}\, x(t) = \int_a^b \varphi(x(t))\, d\ell_g,$$

where ℓ_g is the g-length parameter on the curve. Then the new distance between two points is defined as the infimum of such integrals for all curves joining the points in question.

3.5.C_1. Definition (Floyd, see [Fl]). The boundary $\partial_\varphi X$ is the set of the "new" points of the *metric* completion $\bar{X}_\varphi \supset X$, that is

$$\partial_\varphi X = \bar{X}_\varphi - X.$$

Remarks (a) Unlike the geodesic boundary ∂X which is just a set, the conformal boundary $\partial_\varphi X$ comes with a structure of a *metric* space, for the metric induced from $\bar{X}_\varphi$. This metric on $\partial_\varphi X$ (and on $\bar{X}_\varphi$) will be called the φ-*metric*.

(b) As we want $\partial_\varphi X$ to be non-empty we must choose φ, such that the metric space $(X, \varphi g)$ is *non-complete* . For this we need $\varphi(x)$ which decays sufficiently fast for $x \to \infty$, such that for some curve $x(t)$, $t \in [0, \infty]$, in X with $x(t) \to \infty$, the integral $\int_a^\infty \varphi(x(t))\, d\ell_g$ converges and so $x(t)$ converges in the new metric to some point $x_\infty \in \partial_\varphi X$. On the other hand, we want more than a single point for $\partial_\varphi X$ and so we need a *positive* lower bound for the diameters in the φ-metric of the exteriors of the g-balls $B(r) \subset X$ for $x \to \infty$. Thus $\varphi(x)$ should not decay too fast. We shall see soon that a proper balance can be achieved for manifolds X with $K \le -\kappa^2 < 0$.

3.5.C_2. If two functions φ and φ' are *equivalent* on X in the sense that the ratios φ/φ' and φ'/φ are bounded on X, then the identity map $(X, \varphi^2 g) \leftrightarrow (X, \varphi'^2 g)$ is bi-Lipschitz. Therefore it extends by continuity to a (bi-Lipschitz) homeomorphism of the completions

$$\bar{X}_\varphi \leftrightarrow \bar{X}_{\varphi'}.$$

In particular, $\partial_\varphi X = \partial_\varphi X'$.

More generally, let $f: X \to Y$ be a Lipschitz map and $\varphi(x) = \psi(f(x))$ for some positive function ψ on Y then f is also Lipschitz for the conformal metrics in X and Y with the multipliers φ and ψ. Hence, f extends to a Lipschitz map $\bar{X}_\varphi \to \bar{X}_\psi$ which sends $\partial_\varphi X$ to $\partial_\psi X$ whenever f is *proper*, i.e. $x \to \infty$ implies $f(x) \to \infty$. The same remains true if we replace φ by an equivalent function φ' on X. Furthermore, if f is a bi-Lipschitz homeomorphism the resulting boundary map $\partial_{\varphi'} X \to \partial_\psi Y$ is a homeomorphism.

3.5.C_2'. Example. Let $\psi(y) = (1 + \mathrm{dist}(y_0, y))^\alpha$ for a fixed point $y_0 \in Y$ and some $\alpha \in \mathbb{R}$, and let $\varphi'(x)$ be a similar function with the same exponent α. Then for every bi-Lipschitz map $f: X \to Y$ the composed function $\varphi = \psi \circ f$ obviously is equivalent to φ' and so we obtain the boundary homeomorphism $\partial_{\varphi'} X \to \partial_\psi Y$.

3.5.C_2''. Remark. If X is complete then the boundary $\partial_\varphi X$ is non-empty for the above φ if and only if $\alpha < -1$. Indeed, for $\alpha < -1$ the integral of φ over each geodesic ray in X converges ; on the contrary if $\alpha \geq -1$ one gets divergent integrals over all infinite curves in X.

3.5.D. Let us construct the boundary map $\partial_\varphi X \to \partial_\psi Y$ for an arbitrary quasi-isometry $f: X \dashrightarrow Y$. We assume here that X and Y are complete manifolds and the functions $\varphi(x)$ and $\psi(y)$ decay on X and Y for x and y going to infinity. Furthermore, we assume that $\log \varphi$ and $\log \psi$ are *Lipschitz* which signifies the bound on $\varphi(x_1)/\varphi(x_2)$ in terms of dist (x_1, x_2) and a similar bound for ψ. Finally, we suppose that φ is equivalent to $\psi \circ f$, which means a uniform bound on $\varphi(x)/\psi(y)$ and on $\psi(y)/\varphi(x)$ for all $(x, y) \in X \times Y$ in the graph of f. Notice that all these conditions are satisfied for φ and ψ of the form $(1 + \mathrm{dist})^\alpha$, $\alpha < 0$, as in 3.5.C_2'.

3.5.D'. Proposition. *Under the above assumptions* f *induces a boundary homeomorphism*

$$\partial_\varphi X \to \partial_\psi Y$$

which is, moreover, bi-Lipschitz.

Proof. A straightforward argument shows that our f satisfies the following version of the Lipschitz condition for the metrics $\varphi^2 g$ and $\psi^2 h$ which is somewhat better that the condition (*) in 3.2.B'.

There exists a (Lipschitz) constant $A = A(f, \varphi, \psi)$ *and a function* $b(x_1, x_2)$ *which decays for* $x_1, x_2 \to \infty$ *and* $\mathrm{dist}_{\varphi^2 g}(x_1, x_2) \to 0$ *such that*

$$A^{-1} \mathrm{dist}_{\varphi^2 g}(x_1, x_2) - b \leq \mathrm{dist}_{\psi^2 h}(y_1, y_2) \leq A\, \mathrm{dist}_{\varphi^2 g}(x_1, x_2) + b \qquad (+)$$

for all (x_1, y_1) *and* (x_2, y_2) *in the graph* $\Gamma r_f \subset X \times Y$.

Notice that the improvement here compared to (*) in 3.2.B' is the decay of b, that is

$$b \leq \delta(r, \varepsilon)$$

for $r = \text{dist}_g(x_1, x_0)$ and $\varepsilon = \text{dist}_{\varphi^2 g}(x_1, x_2)$, where $\delta \to 0$ for $r \to \infty$ and $\varepsilon \to 0$. This improvement is mainly due to the decay of φ and ψ.

Now one sees with (+) that f matches the $\varphi^2 g$-Cauchy convergent sequences in X with $\psi^2 h$-Cauchy convergent ones in Y in-so-far as these sequences diverge in the original complete metrics. This matching is exactly what we want,

$$\partial_\varphi X \leftrightarrow \partial_\psi Y .$$

(We suggest to the reader to go through the details more carefully than we have done here).

3.5.D_1'. Example. If φ and ψ are of the type $(1 + \text{dist})^\alpha$ as in 3.4.C_2' then an arbitrary quasi-isometry f : $X \to Y$ induces a homeomorphism $\partial_\varphi X \geq \partial_\psi Y$.

3.5.E. The case $K \leq -\kappa^2 < 0$.

Our discussion on ∂_φ was so far quite general and rather trivial. Now, we assume once again that X is a complete simply connected manifold with $K(X) \leq 0$. We fix a point $x_0 \in X$ and we denote by P_r the *radial projection* of the *exterior* $E(r) \subset X$ of the r-ball, that is

$$E(r) = \{x \in X \mid \text{dist}(x, x_0) \geq r\},$$

to the boundary sphere

$$S(r) = \partial E(r) = \{x \in X \mid \text{dist}\,(x, x_0 \geq r\},$$

Recall that P_r equals the unique intersection point of the geodesic segment $[x_0, x] \subset X$ with S(r).

The convexity of the distance function between (points on) rays issuing from x_0 implies that P_r is a contracting map, that is the differential D of P_r has $\|D\| \leq 1$.

At this point we make a stronger assumption on X, namely, we assume the sectional curvature K(X) is *strictly negative* that is $K(X) \leq -\kappa^2$ where $\kappa > 0$.

For example, X may be a non-compact symmetric space of rank one.

The only feature of $K \leq -\kappa^2$ we need is the following basic bound on the norm of the above D at all points $x \in X$ *outside the sphere* S(r)

$$\|D_x\| \leq \exp - \kappa(r_x - r) \qquad \boxtimes$$

where $r_x = \mathrm{dist}(x, x_o)$ (see [Ch-Eb] for the proof).

Remarks (a) The exponential contraction is usually expressed as the *exponential growth* of the Jacobi fields along the rays issuing from x_o.

(b) We shall later need ⊠ only for a single value r, say for $r = 1$.

(c) Instead of P we could equally use the projection G to the *tangent* sphere at x_o as we did in 3.5.B_1. We prefer here P as it is more geometric and admits a generalization to *singular* spaces of negative curvature (compare [Gr]4).

3.5.E_1. We want to compare the boundary ∂X constructed with geodesic rays (see 3.5) and the conformal boundary $\partial_\varphi X$. If $\varphi(X) \le r_x^{-1-\varepsilon}$, $\varepsilon > 0$ for large r_x, then every geodesic ray R has a finite $\varphi^2 g$-length, that is $\int_R \varphi(r)\,dr < \infty$, and so it defines a point s at $\partial_\varphi X$. If, furthermore, $\|\mathrm{grad}\,\varphi(x)\| \to 0$ for $x \to \infty$, then this s does not change if we take an (equivalent) ray lying within finite Hausdorff distance from R. In this case we obtain a natural map $\partial X \to \partial_\varphi X$ which is clearly, continuous. But unfortunately, this map may easily be constant which is not, of course, what we want.

Yet for $K \le -\kappa^2 < 0$, we are presented with the following surprise

3.5.E_1'. Lemma (See [Fl]). *If $\varphi(x)$ does not decay faster than $\exp - \kappa r_x$ but yet decays at least as $r_x^{-1-\varepsilon}$, then the map $\partial X \to \partial_\varphi X$ is a homeomorphism.*

Proof. Every point $s \in \partial_\varphi X$ can be arrived at by an infinite curve $x(t)$ in X, for $t \in [0, \infty)$, of finite $\varphi^2 g$-length. This means $L_\varphi = \int_0^\infty \varphi(x(t))\,d\ell_g < \infty$ and $\mathrm{dist}_{\varphi^2 g}(s, x(t)) \to 0$ for $t \to \infty$. Given such $x(t)$ we consider the geodesic segments $[x_o, x(t)] \subset X$ and we observe that the decay bound

$$\varphi(x) \ge \mathrm{const}\, \exp - \kappa r_x$$

gives the bound on the length of the radial projection $P(t)$ of $x(t)$ to the unit sphere $S(1) \subset X$ by

$$\text{length } P[0, \infty) \le \mathrm{const}\, L_\varphi < \infty, \qquad \circledast$$

where we assume without loss of generality that $x(t)$ lies in the exterior $E(1) \subset X$ of $S(1)$, which insures the contraction with the rate $\exp - \kappa \operatorname{dist}(x_0, x(t))$. The inequality ⊛ implies the convergence of $P(t)$ to some point in $S(1)$ for $t \to \infty$ and thus the convergence of the rays R_t extending $[x_0, x(t)]$ to a ray R issuing from x_0

Each R_t has finite φ^2g-length and thus defines a point $s_t \in \partial_\varphi X$. This point is joined with s by a double infinite curve C_t which consists of the subray $R_t' \subset R_t$ starting from the point $x(t) \in R_t$ and the part $x[t, \infty)$ of the curve $x[0, \infty) \subset X$,

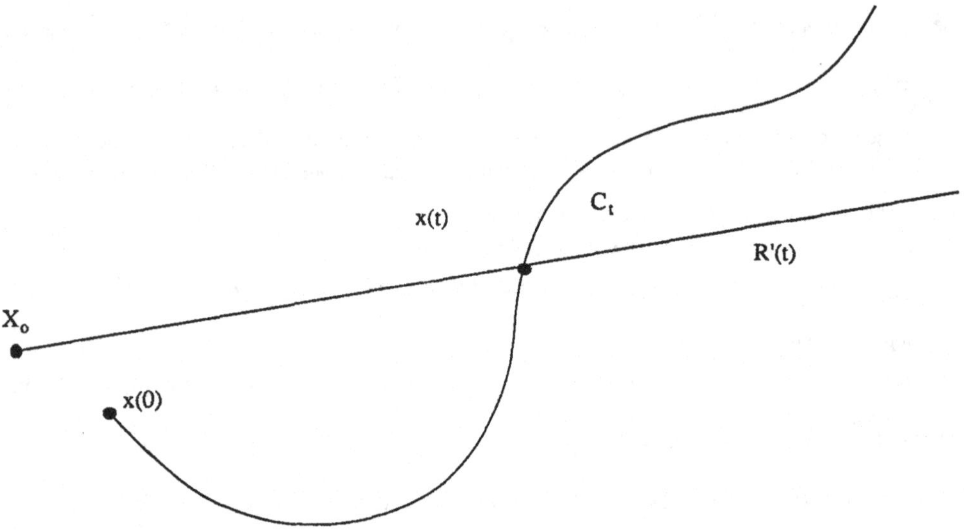

Clearly, $\operatorname{length}_{\varphi^2 g} C_t \to 0$ for $t \to \infty$ and so the limit ray R defines the point $s \in \partial_\varphi X$ we have started with. Therefore the map $\partial X \to \partial_\varphi X$ is surjective.

In order to prove that this map cannot bring together two points represented by different rays R_1 and R_2 issuing from x_0 we must show that the φ^2g-distance $\operatorname{dist}_{\varphi^2 g}(r_1, r_2)$ is bounded away from zero for the points $r_1 \in R_1$ and $r_2 \in R_2$ going to infinity. In fact, let C be an arbitrary curve in X between r_1 and r_2 . If C meets the unit sphere $S(1)$ at x_0, then clearly $\operatorname{length}_{\varphi^2 g} C \geq \operatorname{dist}_{\varphi^2 g} (S(1), S(2)) > 0$, for r_1 and r_2 outside $S(2)$. But if C lies outside $S(1)$ the projection of C to $S(1)$ has

$$\operatorname{length} P(C) \geq \operatorname{dist}(p_1, p_2) > 0$$

for $p_1 = S(1) \cap R_1$ and $p_2 = S(1) \cap R_2$, and so

$$\operatorname{length} C \geq (\mathrm{const})^{-1} \operatorname{length} P(C) > 0$$

as well (compare (*) above).

This concludes the proof of the bijectivity of the map of the geodesic ray boundary ∂X to the conformal boundary $\partial_\varphi X$. Finally, to prove this map is a homeomorphism, one can either invoke Brouwer's principle or check carefully all steps of the proof.

3.5.E_2. Corollary. *Let* X *and* Y *be complete simply connected manifolds with* $K \le -\kappa^2 < 0$. *Then every quasiisomery* $X \dashrightarrow Y$ *entends to a homeomorphism between the geodesic ray boundaries,* $\partial X \to \partial Y$.

Proof. Choose appropriate functions φ and ψ on X and Y, for example $(1 + \mathrm{dist})^\alpha$ with $\alpha = -2$ for the distance functions from fixed points in X and in Y (compare 3.5. C_2'). Then, by the above lemma $\partial_\varphi X = \partial X$ and $\partial_\psi Y = Y$ and then 3.5. D_1'. applies.

Remarks. (a) The above proof is essentially due to Floyd, but the existence of the boundary homeomorphism was discovered earlier by Efremowitz and Tihomirova (see [Ef - Ti]). This homeomorphism also appears in [Mos]1 and [Mar]3 where it is used is the first step in the proof of the Mostow rigidity for lattices $\Gamma \subset \mathrm{Iso}\, X$.

(b) An immediate consequence of the boundary homeomorphism is the invariance of the dimension under quasi-isometries. That is $\dim X = \dim Y$ as the boundaries ∂X and ∂Y are spheres of the dimensions $\dim X - 1$ and $\dim Y - 1$ respectively. Notice that in general (without the assumptions $K < 0$ etc) quasi-isometries may easily change dimension. For example if K is an arbitrary compact manifold, then $X \times K$ is quasi-isometric to X.

(c) The topology of ∂X is not *the only* quasi-isometry invariant of X. For example the real and the complex hyperbolic space $H^{2n}_{\mathbb{R}}$ and $H^{2n}_{\mathbb{C}}$ have homeomorphic boundaries, namely S^{2n-1}, but they are *not quasi-isometric* (compare 3.6.).

(d) The notion of the geodesic ideal boundary ∂X makes sense if X has ordinary boundary, provided this boundary is convex and so the pairs of points in X can be joined by geodesic segments. In fact most of our discussion extends to this case. However, the boundary ∂X in the general case is not homeomorphic to a sphere but may be an arbitrarily complicated finite dimensional compact space. (It is homeomorphic to the projective limit of the spheres $S(i) \subset X$, $i \to \infty$ for the normal projections $S(i+1) \to S(i)$). Moreover, the theory of $K < 0$ extends to a large class of (singular) hyperbolic spaces (see [Gr]4).

3.5.F. The action of the group $\overline{\mathrm{QIs}}$ on ∂X. It is immediate with the *proof* of 3.5.D' *and* 3.5.E_2 that the group $\overline{\mathrm{QIs}}$ acts on ∂X by homeomorphisms. To see this we must check that the homeomorphism on ∂X induced by a quasi-isometry $f : X \dashrightarrow X$ depends only on the equivalence class of f under the relation $|f - f'| < \infty$. But this is obvious as $\partial X = \partial_\varphi X$ with a function $\varphi(x) \to 0$ for $x \to 0$ and so a bounded perturbation f' of f becomes asymptotically zero in the conformal metric $\varphi^2 g$. (Of course all this is built into the proof of 3.5.D'). One also must check that the resulting map

$$\overline{\mathrm{QIs}}\; X \to \mathrm{Homeo}\; \partial X$$

is a homomorphism, but this is again clear from the functionality of the extension of f to $\partial_\varphi X$ (see 3.5.D').

A somewhat less obvious fact here reads.

3.5.F'. *The homomorphism* $\overline{\text{QIs}}\, X \to \text{Homeo}\, \partial X$ *is injective*.

Proof. We need here the following remarkable property of complete simply connected spaces X with $K(X) \le -\kappa^2 < 0$.

$3.5.F_1'$. Morse Lemma. *Let* $Y \subset X$ *be a subset which is quasi-isometric to* $\mathbb{R}$. *Then there is a unique geodesic* Y' *in* X *(which is isometric to* $\mathbb{R}$!), *such that the Hausdorff distance between* X *and* Y' *is finite. Moreover, this distance* δ *is bounded in terms of the constants (A and B in 3.2.B'.) involved in the definition of quasiisometry between* Y *and* $\mathbb{R}$.

The proof can be derived by a direct geometric argument from the contraction property in 3.5.E, but this is not completely trivial. Various versions of the proof can be found in [Mors], [Bus], [Klin], $[\text{Mos}]_4$, $[\text{Gr}]_4$ etc ...

$3.5.F_2'$. Corollary. *Let* f *be a quasi-isometry of* X *which fixes the ideal ends* y_+ *and* y_- *in* ∂X *of a geodesic* Y_0 *in X. Then all points* $y \in Y_0$ *satisfy*

$$\text{dist}(f(y), Y_0) \le \delta < \infty$$

where δ *is bounded in terms of the quasi-isometry constants* (A and B) *of* f.

Proof. The image $f(Y_0)$ is quasiisometric to $Y_0 = \mathbb{R}$ and so there exists a unique geodesic Y' within finite Hausdorff distance δ from $f(Y_0)$. Since f (extended to ∂X) fixes y_+ and y_-, this Y' has y_+ and y_-, as its own ends. It follows (by the convexity of the distance between Y_0 and Y') that Y_0 and Y' have finite Hausdorff distance and by the uniqueness claim of $3.5.F_1'$ the two are, in fact, equal. Q.E.D.

Now we can prove the injectivity of $\overline{\text{QIs}} \to \text{Homeo}$. Take two mutually orthogonal geodesics Y_0 and Y_1 passing through a given point $x \in X$. If a quasi-isometry f fixes ∂X it fixes, in particular, the four ends of our geodesics and so the f-image of x is contained in the intersection I of the δ-neighbourhoods of Y_0 and Y_1. It is easy to see that I is contained in the 2δ-ball around x, and so

$$\text{dist}(x, f(x)) \le 2\delta$$

for all $x \in X$. That is, $f \sim \text{Id}$ as we claimed.

3.5.G. Remark and exercices. (a) The Morse Lemma remains valid for subsets Y quasi-isometric to segments [a, b] and also to $\mathbb{R}_+$ but of course, there is no uniqueness in the case of $\mathbb{R}_+$.

(b) The Morse Lemma for rays gives another proof of the extension of a quasi-isometry $f: X \to Y$ to $\partial X \to \partial Y$. In fact the f-image of every ray R can be δ-approximated by a ray in Y thus giving the corresponding point in ∂Y.

(c) The convergence of $f(R) \subset Y$ to some point in ∂Y can be seen directly without Morse Lemma. For example, if f is a bi-Lipschitz map, then the radial projection of $f(R)$ to the unit sphere $S(1)$ in Y around a fixed point $y_0 \in Y$ has finite length. This immediately implies, (as we have seen already several times) the convergence of this projection to some point $y_1 \in S(1)$ and thus the convergence of $f(R)$ to the point $y_\infty \in \partial X$ represented by the ray extending the segment $[y_0, y_1] \subset Y$.

Notice that the above arguments work for maps more general than quasi-isometries. For example if f is Lipschitz, than the quasi-isometry bound

$$\mathrm{dist}(y_0, f(x)) \geq \mathrm{const}\ \mathrm{dist}(x_0, x)$$

can be easily relaxed to

$$\mathrm{dist}(y_0, f(x)) \geq \mathrm{const}\ \mathrm{dist}(x_0, x)^\alpha, \ \alpha > 0.$$

3.6. Carnot-Mostow metric on the sphere at infinity. Let again X be a complete simply connected manifold with $K \leq 0$ and let S denote the unit tangent sphere at some point $x_0 \in X$.

We consider the radial map of S onto the sphere $S(R) \subset X$ of radius R and we denote by g_R the Riemannian metric on S induced from X. Recall that if $K < -\kappa < 0$, then g_R exponentially grows for $R \to \infty$. In particular

$$\mathrm{Diam}(S, g_R) \geq C^R,$$

for some $C > 1$ and all $R \geq 1$.

Now we denote by di_R the distance function on S corresponding to g_R and normalized by the condition

$$\mathrm{Diam}(S, di_R) = \pi.$$

That is $di_R(s_1, s_2)$ is the g_R length of the shortest path in S between s_1 and s_2 divided by π^{-1} $\mathrm{Diam}_R\,(S, g_R)$.

In general, the family of the functions $di_R : S \times S \to \mathbb{R}$ may diverge for $R \to +\infty$. However one knows that di_R converges in the case where X is a symmetric space. For example, if $\mathrm{rank}X \geq 2$, then the limit metric di_∞ at the sphere $S = \partial X$ equals a positive constant times the Tits metric on ∂X, see [B-G-S]. The Tits metric on ∂X does not depend on the choice of the reference point x^0 and so the action of Iso X on ∂X is di_∞-isometric for $\mathrm{rank}X \geq 2$. The simplest case is that of $X = \mathbb{R}^n$, $n \geq 2$ where (S, di_∞) equals the unit Euclidean sphere S^{n-1} and the group Iso $\mathbb{R}^n$ acts on S^{n-1} via the homomorphism $\mathrm{Iso}\,\mathbb{R}^n \to O(n-1)$.

Now let us describe di_∞ for the symmetric spaces of rank 1.

3.6.A. The space $H^n_{\mathbb{R}}$ of constant curvature -1. Notice that the isotropy subgroup $\mathrm{Iso}_{x_0} \subset \mathrm{Iso}\,\mathbb{H}^n_{\mathbb{R}}$ equals $O(n-1)$ which isometrically acts on the spheres $S(R)$. It follows that $g_R = c(R)\,g_0$ for the unit spherical metric g_0 on S. Therefore di_R equals the distance di_{g_0} associated to g_0 for all $R > 0$ and so $di_\infty = di_{g_0}$ as well. Notice that the conformal factor $c(R)$ equals

$$(\sinh R)^2 = \left(\tfrac{1}{2}(e^R + e^{-R})\right)^2$$

If we change the reference point x_0 then the metric di_∞ also changes and so the action of Iso $\mathbb{H}^n_{\mathbb{R}}$ on (S, di_∞) is *not isometric*. In fact, since this action is faithful (this follows from 3.3.4') and the group $\mathbb{H}^n_{\mathbb{R}} = PO(n, 1)$ is non-compact the boundary action of Iso $\mathbb{H}^n_{\mathbb{R}}$ can not be isometric for any metric on $S = \partial\,\mathbb{H}^n_{\mathbb{R}}$. On the other hand Iso $\mathbb{H}^n_{\mathbb{R}}$ acts on (S, di_∞) by *conformal* transformations. In fact, as everybody knows,

$$\mathrm{Iso} H^n_{\mathbb{R}} = PO(n, 1) = \mathrm{Conf}\, S^{n-1}.$$

3.6.A$_1$. Comparison between the Carnot-Mostow metric and the φ-metric. Let φ be a positive function on $X = H^n_{\mathbb{R}}$ which depends only on dist (x, x_0) (e.g. $\varphi = (\mathrm{dist}(x, x_0) + 1)^{-2}$ as in 3.5.D$_1$.) and let us look at the corresponding φ-metric d_φ on ∂X (compare 3.5.C$_1$.).

Then the $O(n-1)$-symmetry shows that d_φ has the same balls in $\partial X = S^{n-1}$ as the standard metric di_{g_0}. In other words

$$d_\varphi = \Phi(di_{g_0})$$

for some function $\Phi(d)$ depending on φ. If we normalize d_φ by the condition diam $(\partial X, d_\varphi) = \pi$, then, necessarily

$$d_\varphi \le (di_{g_0})$$

because di_{g_0} (unlike d_φ) is a *length* metric associated to a length function on curves. The length property implies (in fact is equivalent to the $O(n-1)$-invariant situation as ours) that di_{g_0} is the *maximal* metric on S^{n-1} which is $O(n-1)$-invariant and having diam $= \pi$.

If the function φ in question has polynomial decay (i.e. $\mathrm{dist}(x_0, x)^\alpha$) then one can show that Φ (d) is of order $|\log d|^{-1}$.

In this situation the Lipschitz property of maps for the φ-metric (established in 3.5.D'.) does not tell much about the di-metric. On the other hand if φ has exponential decay (exp -2 dist) then d_φ and di are Hölder equivalent,

$$d_\varphi \sim (\mathrm{di})^\beta, \ \beta \leq 1$$

as one can prove with a little effort.

We shall explain later on (following Margulis [Mar]3) how the Morse lemma (see 3.5. F_1'.) yields di-*quasiconformality* of the boundary map on ∂X induced by a quasi-isometry of X (alternatively, one could first prove the d_φ-quasiconformality for $\varphi \sim \exp{-\lambda}$ dist and then derive the di-quasiconformality). This quasi-conformality plays the crucial role in Mostow's approach to the rigidity for $H^n_{\mathbb{R}}$ as well as for the other hyperbolic spaces $H^{2n}_{\mathbb{C}}$, $H^{4n}_{\mathbb{H}}$ and $H^{16n}_{\mathbb{C}a}$.

3.6.B. The complex hyperbolic space $H^{2n}_{\mathbb{C}}$. The ideal boundary here is S^{2n-1} which we think of as the unit sphere in $\mathbb{C}^n$. Then there is a natural orthogonal splitting of the tangent bundle.

$$T(S^{2n-1}) = T' \oplus T''$$

where T' is tangent to the Hopf fibers for the S^1-bundle $S^{2n-1} \to \mathbb{C}P^{n-1}$ and T'' is the codimension one subbundle defined with the (standard) complex structure J on the tangent bundle T of $\mathbb{C}^n$ restricted to $S^{2n-1} \subset \mathbb{C}^n$ as follows :

$$T'' = T(S^{2n-1}) \cap J\, T(S^{2n-1}),$$

where $T(S^{2n-1})$ of S^{2n-1} is embedded into T in the obvious way. In other words T'' is the maximal *complex* subbundle in $T(S^{2n-1}) \subset T$.

The splitting $T' \oplus T''$ induces a splitting of the Riemannian metric g_0 on S^{2n-1}, that is

$$g_0 = g_0' + g_0'',$$

where g_0' lives on T' (i.e. vanishes on T'') and g_0'' lives on T''. Now we consider the family $g_\lambda = \lambda g_0' + g_0''$ and try to see what happens for $\lambda \to \infty$.

3.6.B_1. Proposition. *The family of distance* function di_{g_λ} *converges to a certain metric* di *on* S^{2n-1}.

Proof. As metrics di_{g_λ} increase in λ one only has to check they remain bounded, i.e. that

$$\mathrm{Diam}(S^{2n-1}, \mathrm{di}_{g_\lambda}) \leq \mathrm{const}$$

for $\lambda \to \infty$. To prove this it suffices to show that every two points in S^{2n-1} can be joined by a curve C *tangent to* T" (i.e. T(C) $\subset$ T") and having lengthC $\leq$ const. In fact, such a curve has $\text{length}_{g_\lambda} C = \text{length}_{g_0} C$ which implies the bound on Diam.

Now an elementary argument shows that every two points in S^{2n-1} can be joined by a "broken geodesic", that is a piece-wise smooth curve consisting of at most three circular arcs tangent to T". This concludes the proof of the proposition.

3.6.B_1. Remarks (a) It is not hard to see that the limit distance di equals the length of the shortest path tangent to T" between two given points. Metrics of this kind have been discovered by Carnot and Caratheodory (in the framework of the thermodynamical formalism) and have been intensively studied under various names (see [Str]).

(b) The above metric di on S^{2n-1} is highly symmetric. Namely, the natural action of U(n) is di-isometric. Furthermore the boundary action of $U(n, 1) = \text{Iso}(H^{2n}_{\mathbb{C}})$ on $S^{2n-1} = \partial_\infty H^{2n}_{\mathbb{C}}$ is di-conformal as we shall see later on.

3.6.B_2. Now let us look at the spheres $S(R) \subset H^{2n}_{\mathbb{C}}$ around some point $x_0 \in H^{2n}_{\mathbb{C}}$ where the metric in $H^{2n}_{\mathbb{C}}$ is normalized in such a way that the sectional curvature K has inf K = −1. Then the induced metric g_R on $S(R) = S^{2n-1} \subset T_{x_0}(H^{2n}_{\mathbb{C}}) = \mathbb{C}^n$ is given by the formula

$$g_R = e'(R)\, g_0' + e''(R)\, g_0'',$$

where

$$e'(R) = (\sinh r)^2$$

$$e''(R) = (2\sinh r/2)^2$$

To prove this formula for e' we observe that the exponential image of every complex line in $\mathbb{C}^n \supset S^{2n-1}$ (whose intersection with S^{2n-1} is the Hopf fiber where g_0' lives) is a totally geodesic plane (complex geodesic) in $H^{2n}_{\mathbb{C}}$ with constant curvature − 1 (this is easy to prove, compare ([Mos]4) and the e'-formula follows from the corresponding formula for H^2 of curvature − 1.

Now, to understand e", we look at the 2–planes τ in $\mathbb{C}^n = T_{x_0}(H^{2n}_{\mathbb{C}})$ which are *normal* to the Hopf fibers, i.e. $\tau \cap S^{2n-1}$ is tangent to T". Notice that every plane in the real locus $\mathbb{R}^n \subset \mathbb{C}^n$ has this property and, on the other hand, every τ may be brought to $\mathbb{R}^n$ by a unitary transformation of $\mathbb{C}^n$. What is relevant for us in this picture is the existence of τ containing a given vector in $T'' \subset T(S^{2n-1})$. (This follows, for example, from the transitivity of the U(n)-action on such vectors).

Next we recall that the exponential image of $\mathbb{R}^n \subset \mathbb{C}^n = T_{x_0}(H^{2n}_{\mathbb{C}})$ is a totally geodesic submanifold isometric to H^n with curvature $-\frac{1}{4}$ (this is also easy and can be found in [Mos]4). It follows that the metric grows along each τ according to the e"-formula and this formula applies to T" as all vectors there are covered by such planes τ.

3.6.B'_1. Corollary. *The limit (Carnot-Mostow) metric*

$$\mathrm{di}_\infty =_{\mathrm{def}} \lim \pi\, \mathrm{di}_R / \mathrm{Diam}\, \mathrm{di}_R$$

equals the above metric di *on* S^{2n-1} *times the normalizing constant* $= \pi/\mathrm{Diam}\,\mathrm{di}$.

3.6.C. The spaces H\s\up5(4n,ℍ) and H\s\up5(16,ℂa). We use again the Hopf fibration of the sphere at infinity that are $S^{4n-1} \to \mathbb{H}P^{n-1}$ and $S^{15} \to S^8 = \mathbb{C}aP^1$ and we use the *normal* splitting of the tangent bundle of the sphere as in the complex space, that is T' ⊕ T" where T' is tangent to the Hopf fibers. Then we blow up the T'-component of the standard metric in the sphere and arrive as earlier at the metric di. Then it is easy to identify di with the (renormalized) metric di_∞ at the sphere at infinity.

To get some feeling about $\mathrm{di} = \mathrm{di}_\infty$ we mention some properties of this metric.

(a) The Hausdorff dimension of the sphere S with di satisfies (see [Pan]2).

$$\mathrm{Dim}_{\mathrm{Haus}} = \dim_{\mathrm{Top}} S + \mathrm{codim}\, T''.$$

Thus

$$\dim_{\mathrm{Haus}} \partial_\infty(H^n_{\mathbb{R}}) = n-1$$

$$\dim_{\mathrm{Haus}} \partial_\infty(H^{2n}_{\mathbb{C}}) = 2n$$

$$\dim_{\mathrm{Haus}} \partial_\infty(H^{4n}_{\mathbb{C}}) = 4n+2$$

$$\dim_{\mathrm{Haus}} \partial_\infty(H^{16}_{\mathbb{C}}) = 22$$

(b) Let $f: S \to S$ be a C^1-smooth di-Lipschitz map. Then Df maps T" to T" as only the curves *tangent* to T" may have *finite* di-length. This gives no restriction on f for $H^n_{\mathbb{R}}$. In the case of $H^{2n}_{\mathbb{C}}$ the maps of S^{2n-1} preserving the (codimension one) subbundle T" are known under the name of *contact* maps. Such maps form an infinite dimensional Lie group and every contact map is (obviously) di-bi-Lipschitz.

The situation drastically changes for codim T" ≥ 3. Here an old theorem of Cartan says that every *sufficiently smooth* map $S \to S$ preserving T" comes from an isometry of the corresponding hyperbolic space ($H^4_{\mathbb{H}}$ or $H^{16}_{\mathbb{C}a}$). In particular the automorphism group of (S, T") is a (finite dimensional) Lie group. We shall see later on that this conclusion remains valid for all (not necessary smooth) Lipschitz maps and even for di-*quasiconformal* maps.

3.7. Conformal and quasi-conformal maps between metric spaces. Intuitively, a map $f : A \to B$ is *conformal* if it sends "infinitely small balls" in A to such balls in B. To make it precise we introduce the following notion of *asphericity* of a family $\{\mathcal{U}\}$ of neigbourhoods $\mathcal{U} \subset A$ of a fixed point $a \in A$. We denote by inrad $(\mathcal{U}, a)$ the infimum of the distance function dist (a, u) over $u \in \mathcal{U}$ and let outrad denote the supremum of dist(a, u). Then we set

$$\mathrm{asph}(\mathcal{U}, a) = \mathrm{outrad}/\mathrm{inrad}$$

and

$$\mathrm{asph}\{\mathcal{U}\} = \limsup_{\mathrm{Diam}\, \mathcal{U} \to 0} \mathrm{asph}\,(\mathcal{U}, a).$$

For example if all $\mathcal{U}$ are balls around a, then

$$\mathrm{asph}\,\{\mathcal{U}\} = 1.$$

(The number 1 here is the neutral element in the group $\mathbb{R}_+^\times$ which can be turned into 0 by taking log).

Next, for a continuous map $f : A \to B$ we define the (non)-conformality of f at a by

$$\mathrm{conf}_a f = \mathrm{asph}\,\{f(B(r))\}$$

for the balls $B(r) \subset A$ around a for $r \in (0, \infty)$. We call f *conformal* at a if it is conformal at all $a \in A$.

3.7.A. Basic example. Let S be the round sphere with the subbundle $T'' \subset T(S)$ as earlier and g" be the T"-component of the spherical metric. With this g" we have our Carnot-Caratheodory-Mostow metric di on S defined by the length of curves tangent to T".

3.7.A'. Proposition. *A diffeomorphism* $f : S \to S$ *is di-conformal if and only if the differential* Df *sends* T" *to* T" *and the metric* g" *on* T" *goes to a (conformal) metric of the form* $\varphi^2 g''$ *on* T" *for some positive funcion* φ *on* S.

Proof. This is obvious (and known to everybody) in the classical case codim T" = 0 (i.e. T" = T(S) and so g" is a Riemannian metric). In the general case the proof is equally easy once one understands the geometry of di-balls in S of radii $r \to 0$. A little thought shows that such a ball at $s \in S$ looks roughly as a small thickening of a d-dimensional (round) ball $\mathcal{D}(r) \subset S$ for d = rank T" which is tangent at its center to the space T''_s. In other words $\mathcal{D}(r)$ is obtained by exponentiation (for the spherical metric in S) of the r-ball in T''_s. Now, the small thickening $\mathcal{D}_+(r)$ of $\mathcal{D}(r)$ is the ε-neighourhood of $\mathcal{D}(r)$ for $\varepsilon = d^2$. One can show (we suggest it to the reader) that asph $\mathcal{D}_+(r) \to 1$ for $r \to 0$ and then the Proposition follows as in the classical case.

Remark. The above Proposition serves only as an illustration. In our applications of (quasi)-conformal maps to the rigidity we shall use a sligtly different family of quasi-spherical neighbourhoods (see below).

3.7.B. Quasiconformal maps. A continuous map $f: A \to B$ is called *quasi-conformal* if

$$\mathrm{Conf}_a f \leq \lambda < \infty$$

at all points $a \in A$. Sometimes one says f is λ*-quasiconformal* for $\lambda = \sup_{a \in A} \mathrm{conf}_a f$.

3.7.B' Example. Let us look again at maps $f: S \to S$. If such a map is a diffeomorphism then it is quasi-conformal if and only if T'' goes into itself. In fact, the constant λ equals the conformality of the differential Df on T'', i.e. $\sup_{s \in S} \mathrm{Conf}(Df|T''_a)$. The proof follows from that of Proposition 3.7.A'.

3.8. (Quasi) conformal structure on the sphere at infinity. Let us give a different definition of the (quasi)-conformal structure at $S = \partial_\infty X$ for an arbitrary complete simply connected manifold X with $K \leq -\kappa < 0$. We fix a point $x_0 \in X$ and some real number $R_0 > 0$. Then we take a ray Y in X joining x_0 with some point $s \in S$ and we consider the R_0-balls $B(R_0) \subset X$ with centers $y \in Y$. Then we radially project these balls from x_0 to S.

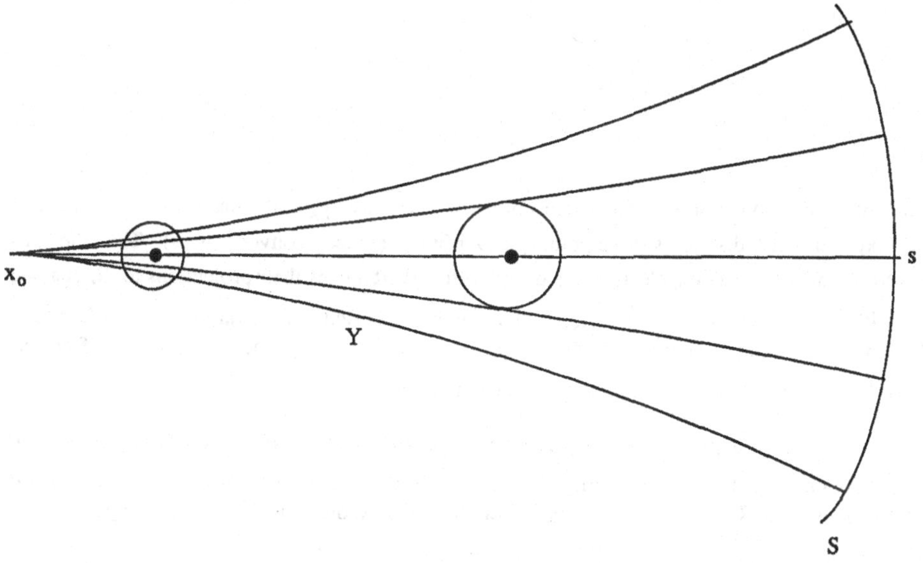

Following Margulis (see [Mar]3) we *call* these projections *balls* in S. More precisely, a Ma-r-*ball*, $r \in (0, 1)$ in S around s is defined as the projection of R_0-ball in X around the point $y \in Y$ with $\mathrm{dist}(x_0, y) = -\log r$. Using these Ma-r-balls one defines in an obvious way the asphericity of (small) neighbourhoods $U \subset S$ and then the rest of the conformal definitions follows.

Examples. If $X = H^n_{\mathbb{R}}$ then the projected balls *are* ordinary spherical balls in $\partial X = S^{n-1}$, but the spherical radius of a Ma-r-ball may be different from r. However,

one can easily see that the spherical radius of a Ma-r-ball is asymptotic (for $r \to 0$) to $C_o r^\alpha$, where α depends on the curvature $K(H^n)$ (if the metric is normalized by $K = -1$, then $\alpha = 1$) and C_o depends on the number R_0 in the Margulis definition. It immediately follows, that these Mar-balls lead to the same (quasi)-conformal geometry : a map $S \to S$ is (quasi) conformal in the usual sense if and only if it is such in Margulis' sense.

In principle, Margulis' structure may depend on x_o and R_o. However, it does not depend on x_o for R_o being kept fixed : the identity map $S \to S$ is conformal for the Mar-conformal structures associated to different points x_o and x_1 in X, provided the curvature of X satisfies $K(X) \le -\kappa < 0$. This follows from the *exponential convergence* at v infinity of the geodesic rays joining x_o and x_1 with a point $s \in S$.

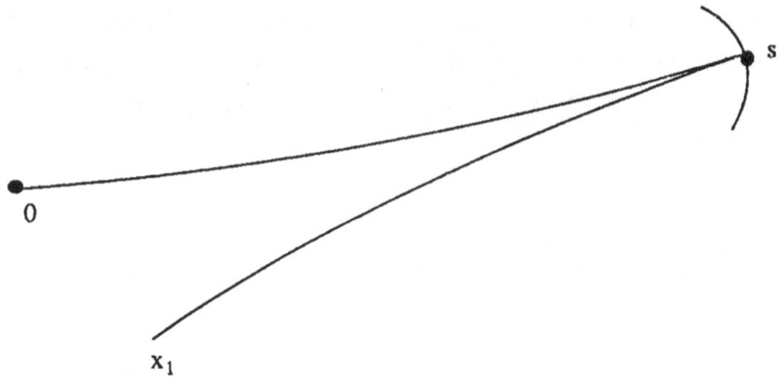

In fact, there is a correspondence between the points of rays, say $y_o \in Y_o$ and $y_1 \in Y_1$, such that $\mathrm{dist}(y_o, y_1)$ (exponentially) decays as these points go to infinity (as they converge to s). Then the balls $B(R_o)$ around y_o and y_1 become closer and closer at infinity. It means their projections to S, that are Ma-r-balls with $r \to 0$ also become very close. Namely the asphericity of the family of the Ma-r-balls projected from x_o with the respect to (the conformal structure defined by) the balls projected from x_1 converges to 1 for $r \to 0$. This is immediate once the definitions are recalled.

Now, if we also change R_o we may change the conformal structure, but the *quasi-conformal* structure does not change. That is the identity map $S \to S$ is quasi-conformal for the structures defined with two different R_o and R_1. This immediately follows from the exponential divergence of rays.

A more profound quasi-conformality result (of Mostow and Margulis see [Mos]4, [Mae]3) concerns the boundary correspondence of a quasi-isometry $X_1 \to X_2$. Here both manifolds X_1 and X_2 are assumed complete simply connected with pinched negative curvature, $0 > -\kappa_1 \ge K \ge -\kappa_2 > -\infty$.

3.8.A. Quasi-conformality theorem. *The boundary map $\partial_\infty X_1 \to \partial_\infty X_2$ induced by a quasi-isometry*

$f: X_1 \to X_2$ *(see 3.2.B'.) is quasiconformal for the Margulis conformal structures in* X_1 *and* X_2.

Sketch of the proof. We may assume the quasi-isometry f in question is full (see 3.3.). Then the images of large R-balls in X have bounded asphericity. Namely if $R > 10B$ for the constant B in the quasi-isometry definition (see(*) in 3.2.B'.). Then the asphericity of f(B) is at most 2A (for A from (*) in 3.2.B'.).

Now we use the Margulis structures with large R_0, we take a small Mar-ball B_1 in $\partial_\infty X_1$, which is the projection of an R_0-ball $B(R_0)$ in X lying far from $x_0 \in X$, and we want to estimate the asphericity of the image $\bar{B}_1 \subset \partial_\infty X_2$ under the boundary map. To do that we invoke the Morse lemma (see 3.5.F. and the discussion which follows) and observe that $\bar{B}_1$ is obtained by a *quasi-radial* projection of $f(B(R_0)) \subset X_2$ to $\partial_\infty X_2$, where "quasi-radial" refers to *quasi-rays* in X_2 which are subsets quasi-isometric to $\mathbb{R}_+$ and which are in our case the f-images of the rays in X_1 projecting $B(R_0)$ to $\partial_\infty X_1$. Denote by $\bar{B}_1'$ the (usual) radial projection of $f(B(R_0))$ to $\partial_\infty X_2$. Now we have *three* structures in $\partial_\infty X_2$. The first one is the quasiconformal structure defined with the Mar-balls B_2 in $\partial_\infty X_2$. Then we have the structure defined with the images $\bar{B}_1$ of the Mar-balls B_1 in $\partial_\infty X$. The third structure is given by the balls $\bar{B}_1'$. The quasi-conformal equivalence between B_2 and $\bar{B}_1'$ structures is obvious with the above asphericity bound for $f(B(R_0)) \subset X_2$. The structures defined with the "balls" $\bar{B}_1'$ and $\bar{B}_1$ are also quasi conformally equivalent but this is less obvious. In fact, it easily follows from the Morse Lemma for quasi-rays. Thus we have the equivalence of B_2-structure to the $\bar{B}_1$-structure which is the image of the Margulis structure in $\partial_\infty X_1$. Q.E.D.

Remark. The above argument is due to Margulis who generalized and simplified the original proof by Mostow in $[\text{Mos}]_1$. A detailed treatment of the classical hyperbolic spaces is given in $[\text{Mos}]_4$.

3.8.B. Let us indicate another more global point of view on the quasi-conformal structure on $\partial_\infty X$ which was much emphasized by D. Sullivan. We use the fact (see [B-G-S]) that every two points s_1 and s_2 in the sphere at infinity $\partial_\infty X$ can be joined by a unique geodesic in X denoted $(s_1, s_2) \subset X$. (This follows, for example, from Morse lemma as the union of two rays $(s_1\ x]$ and $[x, s_2)$ for any $x \in X$ is quasi-geodesic (i.e. quasi-isometric to $\mathbb{R}$ and so can be approximated by a geodesic.

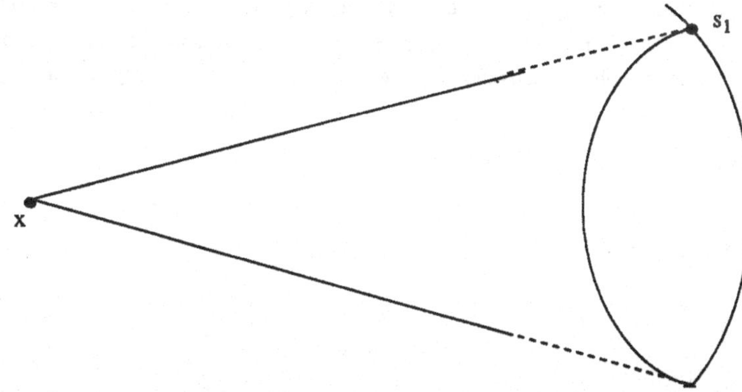

Our conformal structure on $S = \partial_\infty X$ is given by the following function Δ on $S \times S \times S \times S$, $\Delta(s_1, s_2 ; s_3, s_4) = \text{dist}((s_1, s_2), (s_3, s_4))$ where the distance between subsets in X refers to the infimum of the distances between their points. Then a map $S \to S'$ is called quasi-conformal if there exist positive constants a and b such that the functions Δ and Δ' applied to all quadruples of corresponding points s_i and s_i', $i = 1, 2, 3, 4$, satisfy

$$a^{-1}\Delta - b \leq \Delta' \leq a\Delta + b \qquad (+)$$

It is obvious with the Morse lemma that the boundary map of every quasi-isometry is quasiconformal in (+) sense. Conversely, one can show that every (+)-quasi conformal map of the boundary is induced by a quasi-isometry of the underlying space. In other words one can reconstruct X up to quasi isometry from its boundary S. To do that one needs first of all a description of points $x \in X$ in terms of S. The idea is to reverse the map $x \mapsto T_x \subset S \times S$ relating to x the set of T_x the pairs of ends of the geodesic in X going through x. Before doing this we notice the following two properties of $T_x \subset S \times S$.

(i) the projection of T_x on the both components of the Cartesian product $S \times S$ are onto.

(ii) $\qquad \Delta(s_1, s_2 ; s_3, s_4) = 0$ whenever (s_1, s_2) and (s_3, s_4) lie in T_x.

Now, for a given number $\rho > 0$ we consider subsets $T \subset S \times S$ which satisfy (i) and the following modified version of (ii),

(ii)' $\qquad \Delta(s_1, s_2 ; s_3', s_4') \leq \rho$

whenever (s_1, s_2) and (s_3, s_4) lie in T. We call such subsets *ρ-points* and denote by X_ρ the set of all ρ-points $T \subset S \times S$. The distance between two ρ-points T and T' is defined by

$$\sup \Delta(s_1, s_2, s_1', s_2')$$

over all $(s_1, s_2) \in T$ and $(s_1', s_2') \in T'$. It is easy to see that X_ρ with this distance is quasi-isometric to X and that every Δ-quasiconformal map is quasi-isometric on X_ρ if ρ is large compared with the implied constants a and b.

Remark. There is an alternative way to reconstruct X from S based on Ahlfors-Cheeger homeomorphism between the Stiefel bundle Y of the pairs of orthonormal vectors in X and the space of triples of distinct points in X. This homeomorphism maps the pair (τ_1, τ_2) to the ends of the rays defined by τ_1, τ_2 and $-\tau_1$.

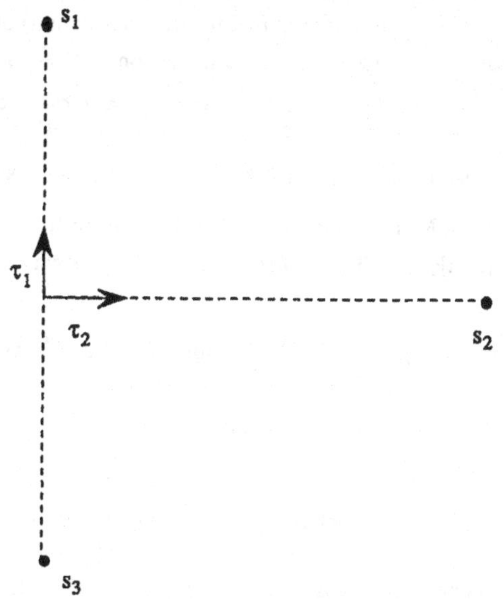

A non-trivial (but not difficult either) point here is bijectivity of this map. Using this we can immediately recapture Y from S and then we observe that the projection $Y \to X$ is a quasi-isometry for the metric structure in Y defined with Δ on S.

3.8.B_1. The quasi-conformal structure of Margulis can be seen using Δ as follows. Take a point s_0 and fix two auxilary points s_1 and s_2 such that $s_0 \neq s_1 \neq s_2 \neq s_0$.

Now let

$$\delta(s) = \Delta(s_0, s\,;\, s_1, s_2),$$

and define Ma-r-balls for $r < 1$ by the condition

$$\{s \in S \mid \delta(s) \geq -\log r\}$$

One can show with no effort at all that the quasiconformal type of this structure at x_0 does not depend on x_1 and x_2. In fact, one can easily see that this structure is equivalent to the Margulis structure.

What looks non-trivial however is the following

<u>Open Problem</u>. *Let* S *be the ideal boundary of a complete simply connected manifold* X *with the curvature pinched between two negative constants.When can one reconstruct the* Δ*-quasiconformal structure from the Margulis structure ?. Equivalently when a Mar-quasi-conformal map* $S \to S$ *is induced by a quasi-isometry of* X*?*

It was been known for many ($\approx$ 50) years that Mar $\Rightarrow \Delta$ for $X = H^n$ and $n \geq 3$. This is equivalent to the existence of *global* quasiconformal invariants (moduli) of subsets in S^{n-1} for $n-1 \geq 2$. On the other hand the local quasiconformal structure on $S^1 = \partial_\infty H^2$ is not strong enough to reconstruct Δ but here one has a refined structure called *quasi-symmetric* (see below 3.10.C_3).

The next known case is that of $S = \partial_\infty H^{2n}_{\mathbb{C}}$ $n \geq 2$ where the global structure can be reconstructed from the local one by the work of M. Reimann [Rei]. A similar (even stronger) conclusion holds for the spaces $H^{4n}_{\mathbb{C}}$ and $H^{16}_{\mathbb{C}}$ (see [Pan]2 and 3.11.A below) but for a somewhat different reason.

3.9. Idea of Mostow's proof of the rigidity for $H^n_{\mathbb{R}}$. We think here of $H^n_{\mathbb{R}}$ as open (Poincaré) ball

$$B^n \subset S^n \subset \mathbb{R}^{n+1}$$

which is formed by the rays in $\mathbb{R}^{n+1}$ on which the quadratic form $h = x_o^2 - \sum_{i=1}^{n} x_i^2$ is positive. Then the group $O(n, 1)$ acts on B^n by conformal transformations (for the usual conformal structure in B^n). Since the isotropy subgroup Iso_x of each point $x \in B^n$ is compact (it is $O(n)$) there exists an $O(n, 1)$ invariant metric g on B^n. This metric is unique up to a scaling constant since the action of Iso_x on $T_x(B^n)$ is irreducible. In fact the metric g turns B^n into the hyperbolic space $H^n_{\mathbb{R}}$ with constant negative curvature whose isometry group equals

$$PO(n, 1) = O(n, 1)/\{\pm 1\},$$

which is the same as the group of conformal automorphisms of the boundary S^{n-1} of B^n. (The ball B naturally embeds into $\mathbb{R}^{n+1}$ as each h-positive ray $b \subset \mathbb{R}^{n+1}$ contains a unique point x, such that $h(x, x) = 1$. The form h is positive on the tangent bundle of B embedded into $\mathbb{R}^{n+1}$ by $b \mapsto x$ and $h|T(B)$ can be taken for our g).

Notice (see the first example in 3.8) that the ordinary conformal structure on S^{n-1} is the same as the Margulis structure. In particular, every quasi-isometry $H^n_{\mathbb{R}} \to H^n_{\mathbb{R}}$ extends to a unique quasiconformal map $S^{n-1} \to S^{n-1}$ for the usual conformal structure of S^n. The basic property of such maps is the following.

3.9.A. Regularity Theorem. *Every quasi-conformal map* $f : S^{n-1} \to S^{n-1}$, $n - 1 \geq 2$, *is almost everywhere differentiable and the differential* Df *is quasiconformal in the following sense. There exists a constant* $\lambda > 0$ *(depending on* f*) such that for almost all* $s \in S^{n-1}$ *and all pairs of unit tangent vectors* τ_1 *and* τ_1 *in* $T_s(S)$ *the norms of their* Df *images satisfy*

$$\lambda^{-1} \|Df(\tau_1)\| \leq \|Df(\tau_2)\| \leq \lambda \|Df(\tau_1)\|.$$

The proof of that can be found in [Vai]. Also notice that the regularity fails for $n - 1 = 1$.

Another ingredient of Mostow's proof is the following special case of Mautner's ergodicity theorem (see [Mau]).

3.9.B. *Let* Γ *be a lattice in* $O(n, 1)$. *Then the action of* Γ *on the projectivised tangent space* $PT(S^{n-1})$ *is ergodic. That is every invariant measurable subset has measure zero or full measure.*

3.9.B'. Corollary. *Let* $f\colon S^{n-1} \to S^{n-1}$ *be a quasi-conformal map such that the action* $s \mapsto f\gamma f^{-1}(s)$ *of* Γ *on* S^{n-1} *is conformal. Then* f *is conformal.*

Proof : Define

$$\varphi(t) = \|Df(\tau)\| / \|Df \mid T_s(S^{n-1})\|$$

for all unit vectors $\tau \in T_s(S)$ and all $s \in S$. As $\varphi(\tau) = \varphi(-\tau)$ this defines a function, also called φ, on $PT(S^{n-1})$. This function clearly is Γ-invariant and by the ergodicity it is a.e. constant. It follows that Df is a.e. conformal, which implies (as one knows, see [Resh], [Geh]) that Df everywhere exists and is conformal. This means f is conformal. Q.E.D.

Now Mostow's proof runs as follows. We consider two compact n-dimensional manifolds V and V' with constant negative curvature and let $\Gamma \to \Gamma'$ be an isomorphism between their fundamental groups. This induces a quasi-isometry between their universal coverings $\tilde{V} \to \tilde{V}'$, both of which are isometric to $H^n_{\mathbb{R}}$, and this quasi-isometry induces a *quasi-conformal* (see 3.8.A.) homeomorphism f between the ideal boundaries of $\tilde{V}$ and $\tilde{V}'$ both of which are conformal to S^{n-1}. The functoriality of the boundary map implies that f carries over the action of Γ on $S^{n-1} = \partial_\infty \tilde{V}$ to the action of Γ' on $S^{n-1} = \partial_\infty \tilde{V}$. Since both actions are conformal, f is conformal and so it extends to a unique isometry between $\tilde{V}$ and $\tilde{V}'$. Since f agrees with the Γ and Γ' actions, same is true for our isometry which shows this isometry comes from an isometry between underlying compact manifolds V and V'. Q.E.D.

3.10. Rigity for $H^{2n}_{\mathbb{C}}$. One can identify $H^{2n}_{\mathbb{C}}$ with the open ball $B^{2n} \subset \mathbb{C}P^n$ which consists of those lines in $\mathbb{C}^{n+1}$ on which the Hermitian form $z_0\bar{z}_0 - \sum_{i=1}^{n} z_i\bar{z}_i$ is positive. Then we have the natural holomorphic action of the group $PU(n, 1)$ on B^{2n} and this action admits a unique up to scale invariant Riemannian metric g which turns B^{2n} into $H^{2n}_{\mathbb{C}}$. The sectional curvatures of g between $-\frac{1}{4}\kappa$ and $-\kappa$ for some $\kappa > 0$ and we normalize g to have $\kappa = 1$.

Now we recall (compare 3.6.B.) the following totally geodesic subspaces in $H^{2n}_{\mathbb{C}}$.

(a) complex geodesics that are traces of projective lines in $\mathbb{C}P^n \supset B^{2n} = H^{2n}_{\mathbb{C}}$. These are isometric to $H^2_{\mathbb{R}}$ with curvature -1.

(b) Isotropic planes, that are $U(n, 1)$-translates of

$$H^2_{\mathbb{R}} \subset H^n_{\mathbb{R}} \subset H^{2n}_{\mathbb{C}} \text{ for } H^n_{\mathbb{R}} = \mathbb{R}P^n \cap B^{2n} = H^{2n}_{\mathbb{C}}$$

for the standard embedding $\mathbb{R}P^n \subset \mathbb{C}P^n$. These planes have constant curvature $-\frac{1}{4}$. The boundary sphere $S^{n-1} = \partial B^{2n} = \partial_\infty H^{2n}_{\mathbb{C}}$ admits a unique $U(n, 1)$-invariant subbundle $T'' \subset T(S^{n-1})$ of codimension one which is the maximal complex subbundle of the real hypersurface $\partial B^{2n} \subset \mathbb{C}P^n$. Furthermore, the *Levi form* of ∂B^{2n} defines a conformal structure (i.e. a positive definite quadratic form up to a multiplication by a positive function on ∂B^{2n}) in T''.

Then the group $PU(n, 1)$ acts on $S^{2n-1} = \partial B^{2n}$ by *conformal transformations* as this group preserves the complex structure as well as the hypersurface ∂B^{2n} and, hence, the conformal class of Levi's form). (Notice that the word "conformal" includes the contact property, that is the preservation of T'' as well as the conformality on T''). Conversely, every conformal transformation of S^{2n-1} is induced by a unique isometry of $H^{2n}_{\mathbb{C}}$, as a (relatively) simple argument (similar to the real case) shows. Notice that the isometry group Iso $H^{2n}_{\mathbb{C}}$ consists of two connected components where the component of Id equals $PU(n, 1)$.

Now, the main step of Mostow rigidity proof is as follows.

3.10.A. *Every quasi-isometry* $H^{2n}_{\mathbb{C}} \to H^{2n}_{\mathbb{C}}$ *induces a Mar-quasiconformal homeomorphism* f *of the boundary* S^{2n-1}.

Here we explain the geometric significance of Mar-quasiconformality. For this we notice that every small Ma-r-ball in S^{2n-1} looks like a small thickening of a $(2n-2)$-dimensional disk tangent at the center to T''. (Compare 3.6.B.) In fact such a ball is obtained by a radial projection of a ball $B(r_0) \subset H^{2n}_{\mathbb{C}}$ to $S^{2n-1} = \partial_\infty H^{2n}_{\mathbb{C}}$. To see that we identify S^{2n-1} with the unit tangent sphere in $T_{x_0}(H^{2n}_{\mathbb{C}})$ and observe that the exponential divergence rate of rays in the complex geodesics passing through x_0 is "twice" as fast as in the isotropic planes. In fact in a complex geodesic (which has curvature -1) the distance between two nearby rays grows as a function of the length parameter t as $\sinh t \sim e^t$ and in the isotropic planes it is $\sim e^{t/2} = \sqrt{e^t}$. It follows, the projection of the R_0 ball to S^{2n-1} is much more squeezed in the complex direction than in the isotropic direction.

$$R_0 \approx \alpha \exp t \approx \beta \exp \frac{1}{2} t.$$

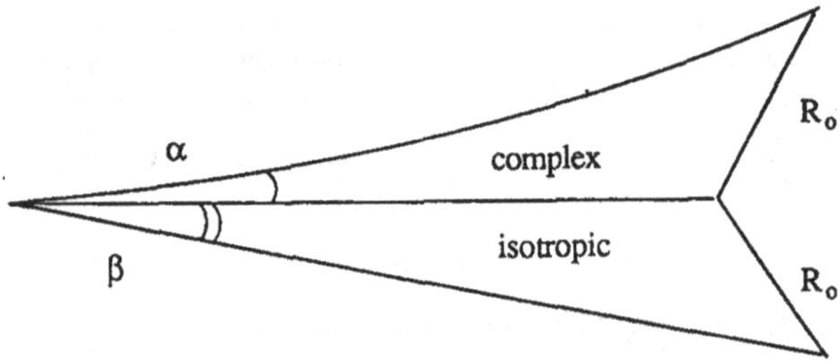

Thus $\alpha \approx \sqrt{\beta}$ which gives us a picture of Ma-r-balls as in 3.6.B. This shows the quasi-conformality of the differential of *smooth* Mar-quasiconformal maps, (which amounts in the smooth case to preservation of T"), but the real issue is to prove this without the smoothness assumption. We need an analogue of Theorem 3.9.A for Mar-quasiconformal mappings. It turns out that even the definition of differentiability has to be adapted to the anisotropic character of the sphere at infinity S of the complex hyperbolic space $H^{2n}_{\mathbb{C}}$. In a sense made precise by Metivier and Mitchell, the "tangent cone" of S at any of its points is a copy of the Heisenberg group. A notion of differentiability can be concocted in such a way that differentials become Heisenberg group automorphisms instead of mere linear maps.

3.10.B. The sphere at infinity of rank one symmetric spaces is modelled on Heisenberg groups (see [Mos]4). We first explain this for $S = \partial X$, $X = H^{2n}_{\mathbb{C}}$. Fix two points s_∞ and s_0 in S (note that the isometry group of X is transitive on the pairs of points of S, so the choice makes no difference). The complement $S - \{s_\infty\}$ should be viewed as a copy of the Heisenberg group $N^{2n-1}_{\mathbb{C}}$.

The Heisenberg group $N^{2n-1}_{\mathbb{C}}$ is the simply connected Lie group with Lie algebra $\mathfrak{N}$ split as

$$\mathfrak{N} = \mathfrak{N}'' \oplus \mathfrak{N}',$$

where

- $\mathfrak{N}'$ is the center, and $\dim \mathfrak{N}' = 1$.
- $\dim \mathfrak{N}'' = 2n-2$ and the Lie bracket

$$[\ ,\ ] : \Lambda^2 \mathfrak{N}'' \to \mathfrak{N}'$$

is a symplectic form.

These properties define a unique group $N^{2n-1}_{\mathbb{C}}$ (up to isomorphism). It possesses a one parameter group of automorphsms δ_t, $t \in \mathbb{R}_+$, defined by

$$\delta_t v = tv \quad \text{for } v \in \mathfrak{N}'',$$

$$\delta_t v = t^2 v \quad \text{for } v \in \mathfrak{N}'.$$

Fix some complex structure J on $\mathfrak{N}''$ compatible with the symplectic structure $[\ ,\ ]$. J determines a metric on $\mathfrak{N}''$. Use left translations to propagate $\mathfrak{N}''$ and its metric into a plane distribution on $N = N^{2n-1}_{\mathbb{C}}$. These data determine a Carnot-Caratheodory metric $\widetilde{di}$ on N (compare 3.6.B_1). This metric is invariant under the left translations, the $U(n-1)$ "rotations", and is multiplied by t under δ_t.

There exists (and unique up to $U(n-1)$) an isometric action of $N^{2n-1}_{\mathbb{C}}$ on $H^{2n}_{\mathbb{C}}$ which fixes s_∞, and has the following properties :

- The map $a : N \to S - \{s_\infty\}$, $n \mapsto n.s_0$ is a diffeomorphism ;
- a takes the left invariant plane field $\mathfrak{N}''$ to T'' ;
- a takes J to the natural complex structure on T'' ;
- a conjugates the group δ_t into a group of isometries of $H^{2n}_{\mathbb{C}}$ which fixes s_∞ and s_0;
- under a, the sphere at infinity of a complex geodesic (compare 3.10) through s_∞ corresponds to an orbit of the action (by translation) of the center ;
- a takes the Carnot-Caratheodory metric $\widetilde{di}$ to a metric conformal, either to the Levi-form (3.10) or the metric di arising as a normalized limit of induced metrics on large spheres (3.6.B_1).

As a consequence, a takes Mar-quasiconformal mappings of S fixing s_∞ to $\widetilde{di}$-quasiconformal mappings of N.

A similar picture exists for the other rank one symmetric spaces. The group $N^{4n-1}_{\mathbb{H}}$ (resp. $N^{15}_{\mathbb{C}a}$) has Lie algebra

$$\mathfrak{N} = \mathfrak{N}'' \oplus \mathfrak{N}'$$

where $\mathfrak{N}''$ identifies with quaternionic space $\mathbb{H}^{n-1}$ (resp. with the Cayley line $\mathbb{C}a$), $\mathfrak{N}'$ with imaginary quaternions $\mathrm{Im}\mathbb{H}$ (resp. imaginary Cayley numbers $\mathrm{Im}\mathbb{C}a$), and the Lie bracket

$$[\ ,\ ] : \Lambda^2 \mathfrak{N}'' \to \mathfrak{N}'$$

is given by

$$\left[(q_1, \dots, q_{n-1}), (q_1', \dots, q_{n-1}')\right] = \mathrm{Im}\left(\sum_{i=1}^{n-1} q_i \overline{q_i'}\right).$$

3.10.C. <u>Differentiability with respect to a group with dilations</u>.

One should think here of the sphere at infinity of the real hyperbolic n-space as modelled on (i.e. locally conformally isomorphic to) $\mathbb{R}^{n-1}$ equipped with its usual dilations. Theorem 3.9.A is concerned with the existence of a differential Df in the usual obvious sense, i.e., if f is a self map of $\mathbb{R}^{n-1}$ and $f(0) = 0$,

$$Df_0 = \lim_{t \to 0} \frac{1}{t} f(th) .$$

This theorem generalizes to groups N equipped with dilations δ_t as we have encountered in 3.10.B.

3.10.C_1. <u>Definition</u> : Let N denote a Heisenberg group over $\mathbb{C}$, $\mathbb{H}$ or $\mathbb{C}a$, and δ_t denote the natural one parameter group of (dilating) automorphisms.

Let f be a map $N \to N$ such that $f(e) = e$. We say that f is *δ-differentiable* at e *with differential* Df_e if, as $t \to 0$, the maps

$$f_t = \delta_{1/t} \circ f \circ \delta_t$$

converge uniformly on compact subsets to Df_e. Using left translations, we define δ-differentiability at any point.

δ-differentiability is somewhat weaker than usual differentiability. Indeed, it essentially means

- f is differentiable in the $\mathcal{H}''$-directions, and Df takes $\mathcal{H}''$ into $\mathcal{H}''$;
- the $\mathcal{H}'$ component of f is differentiable in the $\mathcal{H}'$ direction;
- the $\mathcal{H}'$ component of f behaves like $\sqrt{u}$ under an increment of u in the $\mathcal{H}''$ direction.

Thus a map f which is differentiable in the ordinary sense *and preserves* $\mathcal{H}''$ is not far from being δ-differentiable.

Our main technical result is

3.10.C_2. <u>Regularity theorem</u> : *Every* Mar-*quasiconformal mapping of the sphere at infinity of a rank one symmetric space (locally viewed as a Heisenberg group with dilations) is absolutely continuous, and admits almost everywhere a δ-differential which is a group automorphism commuting with* $\{\delta_t\}$.

The strength of the theorem comes from the assertion that the differential is a group morphism. Indeed, in the complex cases, it states that the derivative in the $\mathcal{H}''$ (or T") direction exists and preserves the natural symplectic form up to a scale. The symplectic form can be viewed as follows : let α be a 1-form on S whose kernel is T" then $d\alpha|_{\mathcal{H}''}$ only depends on $\mathcal{H}''$ (up to a scale). If f is a C^1 diffeomorphism of S which preserves T", then

$$f^*\alpha = \alpha \text{ up to a scale.}$$

The proof that $f^* d\alpha_{|T''} = d\alpha_{|T''}$ up to a scale requires that f is twice differentiable. Thus theorem 3.10.C_2 to some extent asserts existence of some second derivative.

3.10.C_3. There is an unpleasant point here as the existing proofs of theorem 3.10.C_2, in fact, require stronger assumptions on the mapping. Let Y be a metric space. Call the difference set

$$A = B(x, r_2) - B(x, r_1)$$

between concentric balls in Y with ratio of the radii $\frac{r_2}{r_1} = a$ an a-*annulus*. We say that a map between metric spaces $f : Y \to Z$ is *quasisymmetric* if f takes every small enough a-annulus into a η(a)-annulus (where η is some function on $\mathbb{R}_+$).

Quasisymmetry obviously implies quasiconformality and is the only reasonable definition for quasiconformality in one dimension.

The boundary extension of a quasiisometry of $H^{2n}_{\mathbb{C}}$ is easily seen to be quasisymmetric as well as its inverse ; this saves our approach to Mostow rigidity.

Once the a.e. existence and quasiconformality of Df is established the proof is concluded as in the real case with the Mautner lemma which now yields the ergodicity of the Γ-action on PT".

3.11. QIs-*rigidity of* $H^{4n}_{\mathbb{H}}$ *and* $H^{16}_{\mathbb{C}a}$. Recall (from 3.4) that a metric space X is called QI_s-*rigid* if the homomorphism

$$\mathrm{Iso}\,(X) \to \overline{\mathrm{QIs}}\,(X)$$

is bijective.

Let us collect the facts relevant to the QIs-rigidity previously established for X a simply connected Riemannian manifold with curvature $K \le -\kappa < 0$.

a) The natural map $\mathrm{Iso}\,(X) \to \overline{\mathrm{QIs}}\,(X)$ is injective (see 3.3. A').

b) The group $\overline{\mathrm{QIs}}\,(X)$ embeds as a subgroup of homeomorphisms of the sphere at infinity ∂X of X (see 3.5. F'). Furthermore, these homeomorphisms are quasiconformal with respect to the natural Mar-conformal structure on ∂X.

We see that in the context of negative curvature , the QIs-rigidity problem translates into a problem about Mar-quasiconformal mappings of ∂X.

3.11.A. Theorem. : *Let* X *be either quaternionic hyperbolic space* $H^{4n}_{\mathbb{H}}$, $n \ge 2$, *or Cayley hyperbolic plane* $H^{16}_{\mathbb{C}a}$. *Every Mar-quasiconformal mapping of* ∂X *extends to an isometry of* X.

As a consequence, X is $\overline{\mathrm{QIs}}$ -rigid.

3.11.B. The proof follows the lines of Mostow rigidity, as described in 3.9. According to $3.10.C_2$ Mar-quasiconformal (in fact, quasisymmetric, see $3.10.C_3$) mappings of ∂X have almost everywhere differentials which are Heisenberg automorphisms.

The extra rigidity of the quaternionic and Cayley spaces follows from the fact that the corresponding Heisenberg groups have few automorphisms.

Whereas, in the complex case, potential differentials identify with $(2n-2) \times (2n-2)$ matrices that are scalar multiples of symplectic matrices, in the other cases we have

Lemma : *Let* $Z^3 \subset \Lambda^2 H^{n-1}$ (respectively $Z^7 \subset \Lambda^2\, \mathbb{C}a$) *be the* 3-*dimensional subspace generated by the components of*

$$\mathrm{Im}\left(\Sigma\, q_i \bar{q}_i'\right)$$

Let A *be a* $\mathbb{R}$*-linear map of* H^{n-1} *(resp.* $\mathbb{C}a$*) with det* $A = 1$. *If* A *fixes* Z^3 (resp. Z^7) *then* A *is* $\mathbb{H}$*-linear and orthogonal, i.e.,*

$$A \in Sp(n-1),$$

$$(\textit{resp. } A \in Spin(7) \subset Sl(8, \mathbb{R})).$$

Combining $3.11.C_2$ and this lemma, we see that every Mar-quasiconformal mapping of the sphere at infinity of $H^{4n}_{\mathbb{H}}$, $n \geq 2$, or H^{16}_{Ca} is "almost everywhere conformal". The proof is concluded as in the real or complex case.

3.11.C. $H^n_{\mathbb{R}}$ <u>and</u> $H^{2n}_{\mathbb{C}}$ <u>are not</u> $\overline{QIs}$ -<u>rigid</u>.

A selfmapping f ∂X for $X = H^n_{\mathbb{R}}$ or $H^{2n}_{\mathbb{C}}$, has a natural extension $\tilde{f}$ to X in polar coordinates. Indeed, ∂X identifies with the sphere of angular coordinates θ, and one sets

$$\tilde{f}(r, \theta) = (r, f(\theta)).$$

It is readily checked that if f is bilipschitz (case $H^n_{\mathbb{R}}$) and preserves the hyperplane field T" (case $H^{2n}_{\mathbb{C}}$), then $\tilde{f}$ is bilipschitz. Since $\overline{QIs}$ injects into Homeo (∂X), one finds in this way many non trivial elements in $\overline{QIs}$ (X) for $X = H^n_{\mathbb{R}}$ or $H^{2n}_{\mathbb{C}}$. In fact, one has the following

$3.11.C_1$. <u>Theorem</u> (Tukia, Reimann $[Tuk]_2$, [Rei]). *When* $X = H^n_{\mathbb{R}}$ or $H^{2n}_{\mathbb{C}}$, $\overline{QIs}$ (X) *identifies with the group* QC(∂X) *of* Mar-*quasiconformal (resp. quasisymmetric) homeomorphisms of* $S = \partial X$.

The first step in the proof is the fact, alluded to in $3.8.B_1$, that any Mar-quasiconformal mapping of S is Δ-quasiconformal in the sense of 3.8. In fact, the function Δ can be reconstructed in terms of *moduli of curve families* in S or *capacities*, and the point is to show that the relevant capacities are non zero. This is an inequality of Sobolev type.

The second step is simple and general : Δ quasiconformal mappings of $S = \partial X$ extend to quasiisometries of X almost by definition (see 3.8.B.).

$3.11.C_2$. <u>Recapturing</u> O(n, 1) <u>or</u> U(n, 1) <u>from a lattice</u>.

Here is a way to attach a "continuous" group to a discrete group, which in the case of cocompact lattices in G = O(n,1) or U(n, 1) produces G. This method is not functorial under quasiisometries, and thus does not fulfill the program of 3.4.A, but it is functorial under group isomorphisms, and thus implies Mostow-rigidity.

Let Γ be a finitely generated group with a word metric. Say an element f of $\overline{QIs}$ (Γ) is a 1-quasiisometry of Γ if the subgroup generated by Γ in $\overline{QIs}$ (Γ) forms a *uniformly* quasiisometric family (i.e. every transformation φ from this subgroup satisfies the inequality (*) of 3.2.B'. with a constant A independent of φ). We obtain a group $1-\overline{QIs}(\Gamma)$ of 1-quasiisometries.

For Γ a cocompact lattice in O(n, 1), $\overline{QIs}(\Gamma) = QC(S^{n-1})$ identifies with quasiconformal mappings of the standard sphere and a 1-quasiisometry is a mapping of S^{n-1} which generates with Γ a uniformly quasiconformal group. Elaborating on an idea of Mostow, Tukia-Väisälä prove that a uniformly quasiconformal group containing enough conformal transformations is conformal, see [Tuk-Väi].

The argument extends to $H^{2n}_{\mathbb{C}}$, thanks to Reimann's result.

3.12. Topological proof of Mostow rigidity for $H^n_{\mathbb{R}}$.

We adopt the point of view of locally homogeneous spaces explained in 1.5. Then discrete subgroups of the Lie group O(n, 1) correspond to Riemannian manifolds (or orbifolds) with constant curvature -1. Mostow rigidity for cocompact latices in O(n, 1) states.

3.12.A. Theorem : *Let* V, V' *be compact manifolds (orbifolds) with constant curvature* -1. *Then every homotopy equivalence* $V \to V'$ *can be deformed to an isometry.*

We shall again use the fact that homotopy equivalences h between compact manifolds give rise to homeomorphsims f_h of S, the standard sphere viewed as the sphere at infinity of $H^n_{\mathbb{R}}$. The point is to show that f_h has to be a conformal mapping of S (compare 3.9). We shall use a characterization of conformal mappings of S in terms of the regular ideal simplices.

3.12.B. Ideal simplices. Let us identify $H^n_{\mathbb{R}}$ with the ball $B^n \subset S^n$ (see 3.9) and observe that the group O(n, 1) acts on B^n by *projective* transformations which send arcs of great circles in B^n again to such arcs. It follows, that these arcs also serve as the geodesic segments in $H^n_{\mathbb{R}} = B^n$ with an O(n, 1)-equivariant metric g (compare 3.9), and so the projective structure in $H^n_{\mathbb{R}}$ defined by the geodesics is isomorphic to the ordinary projective structure in the ball B^n (which can be equally thought of as a round ball in $\mathbb{R}^n$ rather than in S^n). In particular, one may speak of (geodesically) convex subsets in $H^n_{\mathbb{R}}$ (corresponding to convex subsets in B^n) and of *geodesic simplices* in $H^n_{\mathbb{R}}$ which are *convex hulls* of systems of points in $H^n_{\mathbb{R}}$ in general position. In fact, for every k-tuple of points $x_0, \dots x_k$ in $H^n_{\mathbb{R}}$ one can construct a canonical map of the standard k-simplex Δ^k into $H^n_{\mathbb{R}}$ with the image the convex hull of $\{x_{11}, \dots x_k\}$. Namely, one sends the point with barycentric coordinates $(m_0, \dots, m_k)$ in Δ^k to the *Riemannian center of mass* of the points x_i in $H^n_{\mathbb{R}}$ with the weight m_i assigned to each x_i, $i = 0, 1, \dots, k$. This is the (necessarily unique for $K \le 0$) point $y \in H^n_{\mathbb{R}}$ which minimizes the weighted sum

$$\sum_{i=0}^{k} m_i \operatorname{dist}^2(y, x_i).$$

This map, say $\sigma : \Delta^k \to H^n_{\mathbb{R}}$, is called the *straight* (singular) simplex *spanned by* $x_0, \ldots, x_k$.

Now, let some of the points among $x_0, \ldots, x_k$ lie on the (ideal) boundary $S = H^n_{\mathbb{R}}$ identified with the ordinary boundary of the ball $B = H^n_{\mathbb{R}}$. Then one can still speak of convex hull of points in $H^n_{\mathbb{R}} \cup S = B \cup \partial B$ and then intersect these hulls with $H^n_{\mathbb{R}} \subset H^n_{\mathbb{R}} \cup S$. For example, the convex hull of two distinct points in $S = \partial H^n_{\mathbb{R}}$ is a double infinite geodesic in $H^n_{\mathbb{R}}$. Furthermore, the map $\sigma = \sigma(x_0, \ldots, x_k) : \Delta^k \to H^n_{\mathbb{R}} \cup S$ can also be defined in this case unless among the points $x_0, \ldots x_k$ there are at most two distinct ones. In fact, one defines this σ as the limit (whose existence is easy to prove) of the corresponding maps $\sigma(x_i')$ for $x_i' \in H^n_{\mathbb{R}}$ as these points approximate the points x_i, $i = 0, \ldots, k$. A particularly interesting case is where we have all points $x_1, \ldots, x_k$ on the ideal boundary and then the corresponding map σ is called the *ideal* (straight singular) simplex spanned by these points. Notice that this σ maps Δ^k-{the set of vertices} $\to H^n_{\mathbb{R}}$.

An ideal simplex is called *regular* if it has *maximal symmetry*. This means every permutation of vertices $x_0, \ldots x_k$ in $S = H^n_{\mathbb{R}}$ is induced by some isometry of $H^n_{\mathbb{R}}$, i.e. by a conformal transformation of the sphere $S = S^{n-1}$. Every ideal 2-simplex with distinct vertices is (obviously) regular but it is not at all so for $k \geq 3$. In fact for $n \geq 3$ one has the following elementary

3.12.B_1. Lemma. *A homeomorphism of* S *is conformal if and only if it takes every regular ideal* n*-simplex (here, it is just an* $(n+1)$*-tuple of points in* S) *to another regular simplex.*

Idea of the proof. The case of a general homeomorphism h reduces to that *fixing* some regular $(n+1)$-tuple of points. Then one shows that the preservation of the regularity implies that $h = \mathrm{Id}$. The details are left to the reader.

The following geometric characterization of the regular simplices is also quite elementary but not so easy.

3.12.B_2. Theorem. (Haagerup-Munkholm). *Among all straight* n*-simplices, the regular ideal ones, and only them, have maximal hyperbolic volume.*

See [Haa-Mun] for the proof.

3.12.C. Scheme of the proof of Mostow's rigidity.

Fix an orientation of $H^n_{\mathbb{R}}$ and assume that the manifolds V and V' in question are oriented and the covering maps of $H^n_{\mathbb{R}}$ onto V and V' are orientation preserving. (This can always be achieved by taking the oriented double coverings of V and V' if necessary). The covering maps $H^n_{\mathbb{R}} \to V$ and $H^n_{\mathbb{R}} \to V'$ send singular simplices of $H^n_{\mathbb{R}}$ to those in V and V' and then one naturally defines *straight* simplices in V and V' as well a *straight ideal* and *ideal regular* simplices in V and V'. Then a straight n-simplex (in $H^n_{\mathbb{R}}$, V or in V') is called *positive*, or *positively oriented* if the implied map σ of Δ^n has positive Jacobian for the standard oriented volume element on Δ^n.

Denote by $\mathfrak{R}$ the set of positive ideal simplices in $H^n_{\mathbb{R}}$ and the group of orientation preserving isometries of $H^n_{\mathbb{R}}$ acts transitively on $\mathfrak{R}$. Furthermore if $H^n_{\mathbb{R}}$ is identified with the universal covering $\tilde{V}$ of V then the group $\Gamma = \pi_1(V)$ acts on $\mathfrak{R}$ and this action is cocompact whenever V is compact. As $\mathfrak{R}$ carries a natural $\mathrm{Iso}_+ H^n_{\mathbb{R}}$ -invariant measure, one has a natural measure on $\mathfrak{R}/\Gamma$ and in the cocompact case one can average over this measure. In other words one can average Γ-invariant functions on $\mathfrak{R}$. For example if $f : S \to S$ is the boundary map induced by a given homotopy equivalence $h : V \to V'$ via the covering map $\tilde{h} : \tilde{V} = H^n_{\mathbb{R}} \to H^n_{\mathbb{R}} = \tilde{V}'$, then the function

$$\mathrm{vol}(f\sigma) = \text{volume of } f(\sigma) \subset H^n_{\mathbb{R}}$$

is Γ-equivariant on the set $\mathfrak{R}$ of regular ideal simplices σ in $H^n_{\mathbb{R}}$.

In the next section, we shall explain the following formula

(*) $$\mathrm{average}(\mathrm{vol}(f\sigma)) = \frac{\mathrm{vol}V}{\mathrm{vol}V'}\ \mathrm{average}\ (\mathrm{vol}(\sigma)).$$

We use theorem 3.12.B_2 : for each $\sigma \in \mathfrak{R}$, $f\sigma$ is an ideal simplex, not necessarily regular, so

$$\mathrm{vol}(f\sigma) \le \mathrm{vol}(\sigma).$$

We obtain that

$$\mathrm{vol}(V) \le \mathrm{vol}(V')$$

Reversing h and f, we obtain $\mathrm{vol}(V) = \mathrm{vol}(V')$, thus, for almost every regular ideal simplex σ,

$$\mathrm{vol}(f\sigma) = \mathrm{vol}(\sigma),$$

that is, $f\sigma$ is regular. By continuity, every $f\sigma$ is regular. Lemma 3.12.B, implies that f is conformal. f extends to a Γ-equivariant isometry of $H^n_{\mathbb{R}}$, that is, an isometry of V to V' homotopic to h.

3.12.D. Proof of the formula (*) for average volumes. It stems from a cohomological interpretation of the average. This requires a slight modification of singular homology, in order to admit chains which are compactly supported measures on the space of C^1-singular simplices. Let us admit this modification. Let σ be a singular simplex in $X = H^n_{\mathbb{R}}$ which admits a miror symmetry τ. Let $\pi : X \to V$ be the universal covering map. We define a (generalized) singular chain α in V by

$$a = \int_{\Gamma\backslash G} \operatorname{sign}(g)\, \pi(g\sigma) \frac{dg}{\operatorname{vol} \Gamma\backslash G}$$

where here G is the group of all isometries of X, where $\pm$ sign(g) refers to preservation or reversal of the orientation by $g : X \to X$.

For short, we will denote $\dfrac{dg}{\operatorname{Vol}(\Gamma\backslash G)}$ by $d_\Gamma g$.

3.12.D. Claim : α is a cycle, i.e. $\partial\alpha = 0$

Proof : Recall that, for a simplex σ,

$$\partial\sigma = \sum_{i=0}^{n} \partial_i \sigma$$

where $\partial_i\sigma$ are the (suitably oriented) faces of σ.

We arrange so that the mirror symmetry exchanges the i-th and the (n − i)-th faces :

$$\tau(\partial_i\sigma) = -\partial_{n-i}\sigma.$$

Let τ_i denote the mirror symmetry with respect to $\partial_i\sigma$ then

$$\tau_i(\partial_i\sigma) = -\partial_i\sigma.$$

We compute, using the biinvariance of the Haar measure on $G = \operatorname{Iso} H^n_{\mathbb{R}}$,

$$\begin{aligned}
\partial\alpha &= \sum_{i=0}^{n} \int_{\Gamma\backslash G} \operatorname{sign}(g)\, \pi\, \partial_i(g\sigma)\, d_\Gamma g \\
&= \sum \int \operatorname{sign}(g)\, \pi\, \partial_i(g\tau\, \tau_i\sigma)\, d_\Gamma g \\
&= \sum \int \operatorname{sign}(g)\ \ \pi\, g\tau\, \tau_i(\partial_i\sigma)\, d_\Gamma g \\
&= \sum \int_{\Gamma\backslash G} \operatorname{sign}(g) \left(-\pi g \partial_{n-i}\sigma\right) d_\Gamma g \\
&= -\partial\alpha
\end{aligned}$$

so that $\partial\alpha = 0$.

Thus α defines a homology class $[\alpha]$, which we evaluate against the volume form Ω of V,

$$\langle[\alpha], \Omega\rangle = \int_{\Gamma/G} \text{sign}(g) \left(\int_{\pi(g\sigma)} \text{sign}(g)\ \Omega\right) d_\Gamma g$$

$$= \int_{\Gamma\backslash G} \text{volume}(g\sigma)\ d_\Gamma g$$

$$= \text{volume}(\sigma).$$

Let $h = V \to V'$ be an orientation preserving homotopy equivalence. Let $\pi' : X \to V'$ be the covering map. The image cycle is

$$h_* \alpha = \int_{G\backslash\Gamma} \text{sign}(\sigma)\ \ \pi' \circ h \circ g \circ \sigma\ d_\Gamma g$$

evaluated against the volume form Ω' of V', it gives

$$\langle h_* \alpha, \Omega'\rangle = \int_{\Gamma\backslash G} \text{volume}(h \circ g \circ \sigma)\ d_\Gamma g.$$

Since $[h^* \Omega'] = \frac{\text{vol}\ (V')}{\text{vol}\ (V)} [\Omega]$ in cohomology, we conclude that

$$\int_{\Gamma\backslash G} \text{vol}(h \circ g \circ \sigma)\ d_\Gamma g = \frac{\text{vol}\ V'}{\text{vol}\ V} \int_{\Gamma\backslash G} \text{vol}(g\ \sigma)\ d_\Gamma g.$$

3.12.D_2 Straightening

Assume now that σ is a finite straight simplex in V. We want to replace the curved simplex $h \circ \sigma$ by a straight simplex with the same vertices. This is the straightening operation. This is clearly well defined in $X = H^n_{\mathbb{R}}$ where it is functorial under isometries, and has a functorial chain homotopy

$$1 - \text{straight} = \partial B + B\partial.$$

Then the straightening of simplices in V' covered by X is defined as follows. Given σ in V', take any lift $\tilde{\sigma}$ to X, and set

$$S'(\sigma) = \pi' \text{straight}\ (\tilde{\sigma})$$

Clearly the operator S' induces the identity in homology.

Then we have

$$h_* \alpha = \int_{\Gamma\backslash G} \text{sign}(\sigma)\ \ S'(h \circ g \circ \sigma)\ d g$$

and thus

$$\int_{G/\Gamma} \mathrm{vol}\, S'(h \circ g \circ \sigma)\, d_\Gamma g = \frac{\mathrm{vol}\, V'}{\mathrm{vol}\, V} \int_{\Gamma \backslash G} \mathrm{vol}(g\,\sigma)\, d_\Gamma g\,.$$

Finally, we let σ converge to a positively oriented regular ideal simplex. By continuity, we get formula (*). This finishes the topological proof of Mostow rigidity.

3.12.E. Generalizations.

W. Thurston has extended the above argument to maps $V \to V'$ which are not necessarily homotopy equivalences.

Theorem (W. Thurston). *Let* V, V' *be complete oriented* n-*manifolds of constant curvature* -1 *and let* $h : V \to V'$ *be a proper continuous map of non-zero degree* d. *If* $n \geq 3$ *and* $\mathrm{Vol}\, V \leq |d|\, \mathrm{Vol}\, V' < \infty$ *then* h *is homotopic to a locally isometric* $|d|$-*sheeted covering* $V \to V'$.

The new feature here is a (possible) lack of a *continuous* extension of $\tilde{h} : \tilde{V} \to \tilde{V}'$ to the sphere at infinity. But Thurston (see [Th]) constructs a *measurable* extension using the brownian motion in $\tilde{V}'$.

§ 4 - Rigidity via Bochner formulas for harmonic maps

Harmonic maps f between Riemannian manifolds are the critical points of the *energy functional* E(f) which is the integral of the *energy density* e(f) defined as follows.

4.1. Energy. We use here the norm of linear maps $D : \mathbb{R}^n \to \mathbb{R}^q$ defined by

$$\|D\| = (\text{trace}\, D^*D)^{1/2}$$

Geometrically, we look at the ellipsoid in $\mathbb{R}^q$ which is the image of the in $\mathbb{R}^n$. This ellipsoid has principal semiaxes of certain lengths $\lambda_1 \geq \lambda_2 > \ldots \geq \lambda_q \geq 0$ and

$$\|D\|^2 = \sum_{i=1}^{q} \lambda_i^2 .$$

Another equivalent definition is

$$\|D\|^2 = c_n \int_{S^{n-1}} \|D(s)\|^2 \; ds ,$$

where S^{n-1} denotes the unit sphere in $\mathbb{R}^n$ and

$$c_n = n\,(\text{Vol } S^{n-1})^{-1} .$$

Next for a C^1-map between Riemannian manifolds, say $f : X \to Y$ we denote by $Df : T(X) \to T(Y)$ the differential and by $e(f) = e(f)(x)$ half the squared norm of Df on $T_x(X)$,

$$e(f) = \frac{1}{2} \|Df\|^2$$

This is also called the *pointwise energy* or the *energy density* of f. Then the (global) energy is defined by integration over X,

$$E(f) = \int_X e(f)(x)\, dx$$

One can also obtain E(f) by integrating over the unit tangent bundle S(V) as follows

$$E(f) = c_n \int_S \frac{1}{2} \|Df(s)\|^2 \; ds$$

for the above constant c_n.

Example. If Y is the real line then $E(f) = \int \frac{1}{2} \|\text{grad } f\|^2$ and if $Y = \mathbb{R}^q$ and f is given by the components $f_1, \ldots, f_q$ then

$$E(f) = \sum_{i=1}^{q} \int \frac{1}{2} \|\text{grad } f_i\|^2 .$$

4.2. Non-linear Laplacian. This is a (non-linear differential) operator which assigns to every C^2-map $f: X \to Y$ a *vector field* in Y *along* $f(X)$ which is a loose name for a section of the induced bundle $f^*(T(Y))$ over X. This field is denoted by Δf (it is sometimes called the *tension of* f and denoted by $\tau(f)$) and is defined in several ways as follows.

4.2.A$_1$ Euclidean way. Let first $Y = \mathbb{R}^q$ and $f = (f_1 \dots f_q)$. Then $\Delta f(x)$ is the vector in $T_Y(\mathbb{R}^q) = \mathbb{R}^q$ for $y = f(x)$ with components $\Delta f_1, \Delta f_2, \dots, \Delta f_q$ for the ordinary *Laplace-Beltrami* operator on functions. Then the case of a non-Euclidean Y is reduced to the Euclidean one by using the exponential maps $\exp : T_y(Y) \to Y$ at the points $y = f(x)$ and thus identifying a small neighbourhood of each point $y \in Y$ with $\mathbb{R}^q = T_y(Y)$ for $q = \dim Y$. More precisely, we define $\Delta f(x)$ by first composing f with the inverse exponential at y which gives us a map of (a small neighbourhood of x in) X into $T_y(Y)$. Then we take the above Euclidean Laplacian of this composed map and bring it back to Y with the tautological isomorphism

$$T_0(T_y(Y)) = T_y(Y) .$$

To complete this discussion we recall that the Laplace-Beltrami operator on functions equals the ordinary Laplacian in the *exponential* coordinates,

$$\Delta f(x) = \sum_{i=1}^{n} \partial_i^2 f$$

where ∂_i are the images of coordinate vector fields in $T_x(X) = \mathbb{R}^n$ under the differential of the exponential map. An alternative definition of the Laplace-Beltrami Δ is $\Delta = -d^* d$ where d is the (exterior) differential (thought of as the operator from functions to 1-forms) and d^* is the adjoint operator.

4.2.A$_2$ Laplacian and Hessian. First we recall the second quadratic (fundamental) form II of a submanifold X in a Riemannian manifold Z. This is a symmetric bilinear form on $T(X) \subset T(Z)$ with values in the normal bundle $N(X) \subset T(Z)|X$ defined by

$$\mathrm{II}(\tau_1, \tau_2) = P_N \nabla_{\tau_1} \tau_2$$

where τ_1 and τ_2 are vector fields tangent to X, where ∇ denotes the covariant derivative in Z and P_N is the projection of $T(Z)|X$ to N. One verifies easily that II is indeed a form,

$$\mathrm{II}(\rho_1 \tau_1, \rho_2 \tau_2) = \rho_1 \rho_2 \mathrm{II}(\tau_1, \tau_2) ,$$

for arbitrary functions ρ_1 and ρ_2 on X and it is symmetric, i.e.

$$\mathrm{II}(\tau_1, \tau_2) = \mathrm{II}(\tau_2, \tau_1) .$$

Notice that $X \subset Z$ is *totally geodesic* if and only if $II = 0$.

Next we consider the graph $\Gamma_f \subset W \times Y$ of a C^2-map $f : X \to Y$ and denote by $P : T(X) \to T(\Gamma_f)$ and $Q : N(\Gamma_f) \to f^*(T(Y))$ the obvious maps, where Γ_f is identified with X by

$$(x, f(x)) \leftrightarrow x.$$

Then we define the *Hessian* Hess f by

$$\text{Hess} = Q \circ II \circ P \qquad \text{for } II = II(\Gamma_f).$$

Thus the Hessian is a quadratic form on X with values in $f^*(T(Y))$. Notice that

$$\text{Hess}\, f = 0 \Leftrightarrow II(\Gamma_f) = 0$$

and maps f with Hess $= 0$ are called *geodesic*

Now we set

$$\Delta f = \text{Trace Hess } f.$$

4.2.A_2'. Let us give another description of the Hessian. We denote by α the differential of f viewed as a linear form on X with values in $f^*(T(Y))$. Then, using the covariant derivative ∇^X in $T(X)$ and ∇^* in $f^*(T(Y))$ pulled back from Y, one obtains the covariant derivative $\beta = \nabla\alpha$ which is a symmetric $f^*(T(Y))$-valued bilinear form on $T(X)$ defined by the identity

$$\nabla^*_{\tau_2} \alpha(\tau_1) = \beta(\tau_2, \tau_1) + \alpha(\nabla^X_{\tau_2} \tau_1)$$

for all tangent fields τ_1 and τ_2 on X. In other words,

$$\text{Hess}\, f \underset{\text{def}}{=} \nabla\alpha \underset{\text{def}}{=} \nabla Df$$

and

$$\Delta f = \text{Trace } \nabla\, Df$$

as it should be. (We leave to the reader to check the correctness and the equivalence of the both definitions of Hess).

4.2.A_3. We conclude with a more geometric definition of Δf using the geodesics through $x \in X$. Every such geodesic γ_s defined with a unit tangent vector $s \in T_x(X)$ goes under f into some curve $\tilde{\gamma}_s$ in Y which we parametrize by the length parameter t in X, such that $\tilde{\gamma}_s(0) = y = f(x)$. Then we take the second covariant derivative of $\tilde{\gamma}(t)$ in Y and obtain Δf by averaging on the unit tangent sphere $S^{n-1} \subset T_x(X)$. Namely

$$\Delta f(x) = c_n' \int_{S^{n-1}} \frac{d^2}{dt^2} \tilde{\gamma}_s(0)\, ds,$$

for $c_n' = n/\mathrm{Vol}\, S^{n-1}$. We suggest the reader to prove the equivalence of this to the previous definition in order to obtain certain insight into the meaning of ΔF.

4.2.B. Δf as minus the gradient of E. Let us view the energy $E = E(f)$ as a function on the space of maps $X \to Y$ thought of as an infinite dimensional manifold. The tangent space to this manifold at every f equals the space of fields in Y along $f(X)$ (that are sections $X \to f^*(T(Y))$ and it has a natural scalar product

$$\langle \tau_1, \tau_2 \rangle = \int_X \langle \tau_1, \tau_2 \rangle_Y dx .$$

Then the *Euler-Lagrange equations* for the energy E gives us the gradient of E with respect to the above scalar product. In fact a simple computation shows that :

$$\operatorname{grad} E(f) = -\Delta f, \quad (*)$$

provided X is a compact manifold without boundary. In plain words (*) gives the following formula for the variation of E along one parameter families of maps $f_t : X \to Y$ that are maps $X \times [0, 1] \to Y$.

$$\frac{dE(f_t)}{dt} = -\langle \partial_t f_t, \Delta f_t \rangle, \qquad (**)$$

where $\partial_t f_t = D(\frac{\partial}{\partial t})$ for the differential D of the map $X \times [0, 1] \to Y$ and $\frac{\partial}{\partial t}$ is the coordinate t-field in $X \times [0, 1]$. In fact, (**) is an integral identity

$$\frac{\partial}{\partial t} \int_X e(f_t)(x)\, dx = \int_X \langle \partial_t(f_t)(x), \Delta f_t(x) \rangle_Y dx$$

whose (standard) proof consists in showing that $\delta = \frac{\partial}{\partial t} e(f_t) - \langle .,. \rangle_Y$ is a *divergence* term, that is the divergence of certain function on X.

An important consequence of (*) reads

4.2.B$_1$. *If a* C^2*-smooth map* f *is a critical point for* E *then* $\Delta f = 0$.

Recall that C^2-maps f with $\Delta f = 0$ are called *harmonic*. The above proposition suggests the following approach to the existence of such maps. Start with an arbitrary map $f_0 : X \to Y$ and minimize the energy in the homotopy class $[f_0]$ of f_0. Then, the minimizing map $f \in [f_0]$, provided it exists and is C^2-smooth, is harmonic. We shall see in the following section that this approach is well justified if the receiving manifold Y has non-positive sectional curvature.

4.3. Energy and Laplacian for $K(Y) \leq 0$. Let Y be complete simply connected with $K \leq 0$. Then every two points y_0 and y_1 in Y are joined by a unique geodesic segment $[y_0, y_1] \subset Y$ of points in Y. Notice that this operation for $K(Y) \neq 0$ does not abide the usual rules of commutativity and associativity if the points in question do not lie on a single geodesic. For example, in general

$$\frac{2}{3}\left(\frac{1}{2}x_1+\frac{1}{2}x_2\right)+\frac{1}{3}x_3 \neq \frac{2}{3}\left(\frac{1}{2}x_2+\frac{1}{2}x_3\right)+\frac{1}{3}x_1$$

Next, for every two maps f_0 and f_1 of X to Y one defines the *geodesic* homotopy f_t by

$$f_t(x) = (1-t)\, f_0(x) \ + t\, f_1(x).$$

The convexity of the distance function in Y (insured by $K \leq 0$, see [B-G-S]) implies that the length of each tangent vector $\tau \in T(X)$ in Y is a convex function in t. That is

$$\|Df_t(\tau)\| \leq (1-t)\|Df_0(\tau)\| + t\|Df_1(\tau)\|$$

for all $\tau \in T(X)$ and $t \in [0, 1]$. It follows that the pointwise energy $e(f) = \frac{1}{2}\|Df\|^2$ is convex in t and so the global energy

$$E(f) = \int_X e(f)(x)\, dx$$

is also convex,

$$E((1-t)f_0 + tf_1) \leq (1-t)E(f_0) + tE(f_1) \ .$$

4.3.A. Equivariant maps. Suppose, we are given isometric actions of a group Γ on X and Y and we look at equivariant maps $f : X \to Y$. For example, X and Y may be Galois coverings of compact manifolds V and W and f is a lift of some map $\bar{f} : V \to W$. In general, we assume the action of Γ on X is discrete but me make no such assumption on Y. Since the point-wise energy (obviously) is Γ-invariant on X it descends to a function, also called e on $V = X/\Gamma$ (This may be a singular space if the action is non-free) and we define the Γ-energy by

$$E_\Gamma(f) = \int_V e(v)\, dv\, .$$

In particular, for the above covering example, $E_\Gamma(f)$ equals the energy of the underlying map $\bar{f} : V \to W$.

4.3.A_1. *The Γ-energy is convex for geodesic Γ-equivariant homotopics of maps.*

This is obvious with the convexity of $e(f)$.

Remarks (a). The Γ-energy is most important in our study of the rigidity of groups Γ. Notice that this enery is finite if X/Γ is compact while the ordinary energy (obtained by the integration over X rather than X/Γ) is usually infinite for Γ-invariant maps.

(b) Let us look again at Γ-equivariant maps $X \to Y$ covering some maps $V \to W$. In this case $\pi_1(W) = \Gamma$ and Y is the universal covering of W. Then every continuous map $V \to W$ induces a Γ-covering X of V along with a Γ-invariant map $X \to Y$. This Γ-space X depends only on the homotopy class of the map $V \to W$ and homotopies of maps $V \to W$ correspond to Γ-equivariant homotopies of maps $X \to Y$. Notice that there is a slight ambiguity in there as the equality $\pi_1(W) = \Gamma$ needs a reference point $w_0 \in W$ but this causes no problem.

(c) If the action of Γ on Y is non-discrete we cannot go to the quotient space $W = Y/\Gamma$. Yet, if the action on X is free and discrete we can work with $V = X/\Gamma$ and the fibration $\bar{Z} \to V$, where $\bar{Z} = X \times Y/\Gamma$ for the diagonal action of Γ on $X \times Y$. This is the fibration with the fiber Y associated to the principal Γ-fibration (covering) $X \to V$. Now, Γ-equivariant maps correspond to sections $V \to \bar{Z}$. Every section $\bar{f} : V \to \bar{Z}$ has its point-wise energy $e(\bar{f})$ since, locally, $\bar{f}$ defines a map $V \to Y$ corresponding to the projection $X \times Y \to Y$. Then we have the global energy

$$E(\bar{f}) = \int_V e(\bar{f}).$$

which clearly equals the Γ-energy of the corresponding map $f : X \to Y$.

4.3.A_2. The basic relation between harmonicity and energy remain valid for equivariant maps. Namely, if *some smooth* Γ*-equivariant map* $f : X \to Y$ *gives minimum to* E_Γ *then* f *is harmonic*.

The proof is the same as in the absolute case. Notice that for the covering example the above harmonic maps corresponds to harmonic maps $V \to W$ minimizing the energy in a given homotopy class. In the general case we have harmonic sections $V \to \bar{Z}$ which are sometimes called *twisted harmonic maps.*

4.4. <u>Uniqueness of harmonic maps</u>. The convexity of the energy shows that the map f giving the minimum to $E(f)$ is essentially unique. Before giving a precise statement we look at several examples.

4.4.A. let $V = S^1$. Then harmonic maps $S^1 \to W$ are just parametrized closed geodesics in W. If $K(W) < 0$ then there is a unique geodesic in every (non-trivial) homotopy class. (For the trivial class the harmonic maps are constant map $S^1 \to w \in W$. There is no uniqueness here as w may be any point in W). The only non-uniqueness here comes from the action of S^1 on this geodesic.

If $K(W) \le 0$, then the above uniqueness may fail. For example, if $W = W_0 \times S^1$, then the geodesic maps homotopic to $w_0 \times S^1$ are parametrized by the points $w \in W$. Namely, there is a (unique) harmonic map $S^1 \to W$ in this homotopy class sending a given point $s \in S^1$ to w.

More generally, it may happen that W contains a totally geodesic submanifold isometric to $W_0 \times S^1$. Then one has a family of geodesics in W parametrized by this submanifold.

Now for any V, we may start with a harmonic map f of V into some W_1 and then we have many harmonic maps into $W_0 \times W_1$, namely the maps $f_w(v) = (w, f(v))$ for each point $w \in W_0$. Furthermore, we may have such a product inside of a larger manifold W and then the same non-uniqueness persists for maps $V \to W$.

4.4.B. The following theorem shows that the above examples essentially exhaust all possibilities of non-unique harmonic maps in the case where $V = X/\Gamma$ is compact. To simplify the matter, we assume the manifold Y is real analytic and suppose there exist two Γ-homotopic Γ-equivariant harmonic maps f_0 and $f_1 : X \to Y$, where X is connected and X/Γ is compact.

4.4.B'. Theorem. *The images of the maps* f_0 *and* f_1 *lie in a complete totally geodesic submanifold* $Y' \subset Y$, *such that* Y' *isometrically splits by* $Y' = Y_0 \times \mathbb{R}$ *and* $f_1 : X \to Y'$ *is obtained from* f_0 *by an* $\mathbb{R}$*-translation of* Y' *that is the map* $(y_0, t) \mapsto (y_0, t + t_0)$ *for some* $t_0 \in \mathbb{R}$.

This result is well known and easily follows from the elementary geometry of $K \le 0$ (see [B-G-S]).

Remark. Notice that we do not assume our maps are energy minimizing. In fact one can easily show that harmonic maps are necessarily energy minimizing if $V = X/\Gamma$ is compact.

4.5. On the existence of harmonic maps. A natural approach of constructing a harmonic maps $f : V \to W$ in a given homotopy class $\mathfrak{F}$ consists in taking an energy minimizing sequence of maps $f_i \in \mathfrak{F}$ and taking a sublimit of f_i for f. (Recall that "minimizing" means $E(f_i) \to \inf_{f \in \mathfrak{F}} E(f)$ and "sublimit" is the limit of a subsequence). Let us indicate some examples showing what can go wrong.

4.5.A_1. Let $V = W = S^n$ and $f_t : W \to W$ be the north pole south pole tranformations. Namely f_t fix the north pole $s_+ \in S^n$ and act on $\mathbb{R}^n = S - \{s_+\}$ by $s \mapsto e^t s$, where $\mathbb{R}^n$ is identified with $S^n - \{s_+\}$ by the stereographic projection. The maps f_t are *conformal* and a trivial computation shows that

(1) if $n = 1$ then $E(f_t) \to \infty$ for $t \to \infty$.

(2) if $n = 2$ then $E(f_t) = E(f_0)$ for all t

(3) if $n = 3$ then $E(f_t) \to 0$ if $t \to \infty$.

This example indicates invalidity of the variational approach as the maps f_t for $t \to +\infty$ converge almost everywhere (in fact, everywhere except at the south pole $= 0 \in \mathbb{R}^n$) to the constant map $S^n \to s_+ \in S^n$, which is (though harmonic) not homotopic to f_t. (One sees a more instructive picture by looking at the graphs of f_t that are maps $S^n \to S^n \times S^n$ which are harmonic for $n = 2$).

Also notice that the family is pointwise C^1-*divergent* as the energy density $e(f_t) = \frac{1}{2}\|Df_t\|^2$ blows up at the south pole.

4.5.A_2. The discontinuity phenomenon of the above example can not happen for $K(W) \le 0$ (see below) but one can have another kind of degeneration, where the minimizing maps run to infinity all-to-gether. To see that let $W = H^2_{\mathbb{R}}/\mathbb{Z}$ be the standard *cusp*, that is the group $\mathbb{Z} \subset \mathrm{Iso}\, H^2_{\mathbb{R}}$ consists of *parabolic* elements. This means, there is a *single* point $s \in S_\infty(H^2_{\mathbb{R}})$ fixed under this group. Topologically, $W = S^1 \times \mathbb{R}$ and the metric is $e^{-t}ds^2 + dt^2$.

Now we look at the maps $f_t : S^1 \to W$ given by $f_t(s) = (s, t)$ and observe that $E(f_t) \to 0$ for $t \to \infty$. In fact, even the pointwise energy $e(f_t) = \frac{1}{2}\|Df_t\|^2$ exponentially decays as $t \to \infty$. In this case no limit (or sublimit) of f_t exists and there is no harmonic map in the homotopy class of f_t

4.5.B. Smoothing . In order to prevent a blow up of $e(f)$ (as in 4.5 A_1) one may try to regularize a given minimizing sequence f_i by applying some smoothing operators to f_i. The smoothed maps f_i^{sm} must have roughly the same energy in order to have

$$\lim_{i \to \infty} E(f_i^{sm}) = \lim_{i \to \infty} E(f_i),$$

and the energy density should be uniformly bounded,

$$\sup_{v \in V} e\,(f_i^{sm})\,(v) \le \text{const.}$$

Let us explain why the condition $K(W) \le 0$ makes the existence of a smoothing possible.

4.5.B_1. The center of mass. Let Y be a complete simply connected manifold with $K(Y) \le 0$ (it will be the universal covering of W) and μ be a finite measure on Y. Then we define the *center* of μ, denoted $\bar{\mu}$ or $\int_Y d\mu$ as the (unique !) minimum point of the function d^2_μ, $x \in Y$, obtained by integrating away the second argument in $\mathrm{dist}^2(x, y)$ with respect to μ. Notice, that because of $K \le 0$ each function $d^2_y(x) = \mathrm{dist}^2(x, y)$ is strictly (geodesically) convex on y and so the same is true for d^2_μ. This insures the uniqueness of the minimum point of $d^2_\mu(x)$.

Example. Let μ be supported at two points y_0 and y_1 in Y with some weights p_0 and $p_1 = 1 - p_0$. Then $\bar{\mu}$ equals the convex combination $p_0 y_0 + p_1 y_1$ as defined in 4.3.

4.5.B_2. Averaging families of maps. Let $f_t : X \to Y$, be a family of maps where t runs over some space T with a given finite measure ν on T. Then for every $x \in X$ we take the push-forward measure μ_x for the map $t \mapsto (f_t)\, f_t(x) \in Y$ and then take the center $\bar{\mu}_x \in Y$ for every $x \in X$. Thus we get the averaged map $f = \bar{f}_t : X \to Y$.

The convexity properties of Y with $K(Y) \leq 0$ (which we have met already several times) show that the energy density e(f) at each point $x \in X$ is bounded by the μ-average of the energies of f_t at x. This allows the smoothing of maps into Y in the following.

Flat example. Let V be the flat n-dimensional torus and μ_v be the Lebesgue measure of a ball $B_v(\varepsilon) \subset V$ around $v \in V$ of a small radius $\varepsilon > 0$ such that this $B_v(\varepsilon)$ is isometric to the Euclidean ball $B_0(\varepsilon) \subset \mathbb{R}^n$. Then for every map $f : V \to W$ we have the *smoothed map* $f^\varepsilon : V \to W$ defined by the averaging the push forward of these measures, that is

$$f^\varepsilon(v) = \overline{f_*(\mu_v)},$$

where the averaging is obtained by first lifting f to the universal covering $Y \to W$ and then projecting back to Y. Let us show this smoothing does what we want. Namely

(A) $E(f^\varepsilon) \leq E(f)$

(B) $\sup_{v \in V} e(f^\varepsilon(v)) \leq \text{const}_\varepsilon\, E(f)$

Proof. Let $B(\varepsilon) \subset V = T^n$ be the ε-ball around the identity element in the torus thought of as an additive group. Then we consider the maps $f_t(v) = f(v + t)$ for $v \in V$ and $t \in B(\varepsilon)$ and let μ be the Lebesgue measures on $B_0(\varepsilon)$. Clearly,

$$\bar{f}_t = f^\varepsilon$$

and then (A) and (B) immediatly fall through.

4.5.C. The heat flow. The idea of the extension of the above to non-flat manifolds V consists in taking an "infinitely small" ε and performing the smoothing procedure "infinitely many" times. To keep track of what happens it is easier to replace the measures on the ε-balls by the heat flow on V. Namely, we denote by $\mu(v, t)$ the result of the diffusion of the δ-measure $\delta(v)$ on V at the moment t. In other words $\mu(v, t)$ is the measure on V whose density $\varphi(v')$ is the heat kernel $h(v, v', t)$ that is the solution of the heat equation on V,

$$\Delta_{v'} h = \frac{dh}{dt}$$

with the initial condition

$$h(v', 0) = \delta(v).$$

For example if $V = \mathbb{R}^n$ then

$$h(v, v', t) = (4\pi t)^{-\frac{n}{2}} e^{-\frac{|v-v'|^2}{2t}} .$$

Recall that the *heat flow* defined by h has the following *semigroup property* schematically expressed by

$$h(t_1) * h(t_2) = h(t_1 + t_2)$$

and saying in effect that the measure $\mu(v, t_1 + t_2)$ is obtained by t_2-diffusing each point $v \in V$ with the weight $\mu(v, t_1)$. Namely

$$h\left(v, v', t_1 + t_2\right) = \int h\left(v, v'', t_1\right) h\left(v'', v', t_2\right) dv''$$

Now for maps $f: V \to W$ we define the smoothing operator $H_\varepsilon f$ by

$$H_\varepsilon f(v) = \bar{f}_*(\mu(v, \varepsilon))$$

and then for a given t we apply this operator n-times for $\varepsilon = t/n$, and get

$$H_\varepsilon^n f = H_\varepsilon(H_\varepsilon \ldots (H_\varepsilon f) \ldots) .$$

(Here, as earlier, the averaging refers to the universal covering Y of W). The following result is due to Eells and Sampson (see [Eel-Sam]).

4.5.C_1. Theorem. *If* V *is compact and* $K(W) \le 0$, *then for every smooth map* f *the sequence* $H_\varepsilon^n f$ *converges to a smooth map, denoted*

$$f(v, t) = \mathcal{H}_t(f),$$

and called the (non-linear) heat flow or diffusion of $f = f(v, 0)$. *Furthermore, this flow satisfies the heat equation*

$$\frac{df(v,t)}{dt} = \Delta f(v, t)$$

and the initial condition

$$f(v, 0) = f(v)$$

4.5.C_1'. Remarks (a) As we know (*) says that f(v, t) is the minus gradient flow for the energy and so E(f (v, t)) goes down with $t \to \infty$.

(b) *The theorem remains valid for a* Γ*-equivariant map* $f: X \to Y$, *provided* X/Γ *is compact.*

4.5.D. Bochner's formula and apriori estimation. A standard method of proving the existence of regular solutions of P.D.E. is by first establishing a conditional result giving a bound on certain derivatives of solutions *assuming* a solution exists. Let us indicate such a bound for harmonic maps (and later on for the heat flow) $V \to W$. The basic tool here is the following Bochner formula of Eells and Sampson which holds true for all C^2-smooth harmonic maps $f : V \to W$,

$$\Delta e(f) = \|\mathrm{Hess} f\|^2 + Q_V - Q_W \qquad (*)$$

where Q_V and Q_W are the following expressions involving Df and the curvatures of V and W. First

$$Q_V = \mathrm{Ricci}(Df, Df),$$

where the Ricci tensor in V is thought of as a quadratic form on T(V) which naturally defines with the Riemannian metric in W a quadratic form on the bundle $\mathrm{Hom}(T(V), f^*T(W))$.

Then this extended Ricci applies to $Df \in \mathrm{Hom}(T(V), f^*T(W))$ and defines Q_V.

Now, to define Q_W we take a frame of orthonormal vectors $\tau_1 \dots \tau_n$ at every point v in V and we set

$$Q_W(v) = \sum_{i,j,k,\ell} \left\langle R\left(D\tau_i, D\tau_j\right) D\tau_k, D\tau_\ell \right\rangle_W,$$

where R denotes the curvature tensor of W at $w = f(v)$ and $D = Df$

Notice that $Q_W \le 0$ if $K(W) \le 0$.

We don't prove (*) (see [Eel-Sam]) but rather explain its meaning. First of all, if V and W are flat, then the formula reduces to

$$\Delta e(f) = \|\mathrm{Hess} f\|^2 \qquad (*_0)$$

which is a well known elementary (Euclidean) identity. The major consequence of this is the inequality

$$\Delta e(f) \ge 0 \qquad (*+)$$

which says that the energy density is a subharmonic function on V.

Another important corollary of $(*_0)$ for *closed* manifold V reads

$$\int_V \|\mathrm{Hess}\, f\|^2 = 0 \qquad (*\mathrm{i})$$

Hence $\mathrm{Hess}\, f = 0$ and so *every harmonic map is geodesic*.

Now if V is flat, then the condition $K(W) \leq 0$ reinforces (∗ +) as the term $-Q_W$ in (∗) is positive. The same happen to (∗ i) and so we see that *every harmonic map of a compact flat manifold* V *into* W *with* $K(W) \leq 0$ *is geodesic.*

This can be also seen with the averaging on V which works for all flat manifolds as well as for tori (see 4.5.B_2). The averaged map satisfies $E(f^\epsilon) < E(f)$ unless f is a geodesic map, as a simple convexity argument shows (compare 4.4.B'). Thus every energy minimizing (harmonic) map must be geodesic.

Now we allow non-flat V and observe that $K(W) \leq 0$ implies via (*) the following inequality on $e(f) = \|Df\|^2$,

$$\Delta e(v) \geq Q_V \geq -C(v)\, e(v) \qquad (+)$$

where $-C(v)$ is the lower bound for Ricci_V at v. If V is compact (+) implies the bound

$$\sup_{v \in V} e(v) \leq \mathrm{const}\, E(f) = \int_V e(v)\, dv .$$

Then this point-wise bound on $\|Df\|$ can be extended with little extra effort to the higher order derivatives (see [Eel-Sam]).

4.5.D_1. Bochner for the heat flow. Now let f(v,t) be the solution to the heat equation $\frac{df}{dt} = \Delta f$. Then the Bochner formula generalizes to

$$\Delta e(v) = \|\mathrm{Hess} f\|^2 + Q_V - Q_W + \frac{de(f)}{dt} \qquad (**)$$

where f is viewed as function in v with t as a parameter.

This implies, for $K(W) \leq 0$ in (+) that

$$\frac{de}{dt} \leq \Delta e + C(v) e$$

which leads by a simple argument to the following important conclusion.

4.5.D_1'. *If* V *is compact then the map* f(v, t) *satisfies for every* $t \geq 1$ *the following point-wise bound on* $e(f) = \frac{1}{2}\|Df\|^2$,

$$e(v,t) \leq \mathrm{const}\, Ef(v,0),$$

where

$$Ef(v,0) = \int_V e(v,0)\, dv ,$$

for $e = e(f)$. *Furthermore, a similar bound remains valid for* Γ*-invariant maps* $X \to Y$ *if* $V = X/\Gamma$ *is compact.*

This shows that the diffusion (heat flow) operator $f \mapsto \mathcal{H}_t(f)$ (compare 4.5.C_1) for each $t \geq 1$ does smooth maps $f : V \to W$. Moreover, one can prove that all derivatives of $f(v, t) = \mathcal{H}_t(f)$ are controlled by $E(f)$ and so $\mathcal{H}_t$ provides us with a perfect smoothing. Besides $\mathcal{H}$ *decreases* the energy of maps (see 4.5.C_1') and so it is well adapted to the minimization process. In fact one can use $\mathcal{H}_t(f) = f(v, t)$ as a minimizing family. Namely, the energy $E\mathcal{H}_t(f)$ converges to $\operatorname{Inf} E(f)$ over all f in a given homotopy class and $\Delta\, \mathcal{H}_t(f) \to 0$ for $t \to \infty$. This is shown with yet another Bochner formula, this time for

$$k = \|\Delta f\|^2 = \left\|\frac{df}{dt}\right\|^2$$

satisfied by $f(v, t) = \mathcal{H}_t\, f(v)$.

This formula shows that (see [Eel-Lem]$_1$, p. 24)

$$\frac{dk}{dt} \leq \Delta k \qquad (\square)$$

which immediately implies for compact V the asymptotic bound

$$\lim_{t \to \infty} k(v, t) \leq K(t_0) = \int_V k(v, t_0)\, dv ,$$

for every fixed t_0. Then by integrating ($\square$) over W we see that $K(t)$ is decreasing in t. On the other hand

$$K(t) = -\frac{dEf(v,t)}{dt} ,$$

since $\Delta f = -\operatorname{grad} Ef$. It follows that $K(t) \to 0$ for $t \to \infty$, because Ef remains positive for all t, and so $k(v, t) \to 0$ as well. Thus $\Delta f(v, t) \to 0$ for $t \to \infty$. Since all derivatives of $f(v, t)$ are bounded for $t \to \infty$ there are only two alternatives .

(A) The maps $f(v, t)$ subconverge for $t \to \infty$ to a harmonic map f in the homotopy class of $f(v, 0)$

(B) The maps f move $f(V) \subset W$ away to infinity for $t \to \infty$. That is

$$\operatorname*{Inf}_{v \in V} \operatorname{dist}\left(w_0, f(v, t)\right) \underset{t \to \infty}{\longrightarrow} \infty ,$$

for a fixed point $w \in W$. (Compare example 4.5.A_2).

Notice that in the course of this movement the norm $\|Df\|$ remains uniformly bounded. This implies the uniform bound on the length of loops representing the image $f_*(\pi_1(V)) \subset \pi_1(W)$ as these loops move to infinity together with $f(V) \subset W$.

Example. (See [Don]). Let $K(W) \leq -\kappa < 0$. Then in the case (B) the group $f_*(\pi_1(V)) \subset \pi_1(W)$ acting on the universal covering Y of W is *parabolic* : it fixes a unique point on the ideal boundary of W. In particular if

$$-\infty < -\kappa' \leq K(W) \leq -\kappa < 0,$$

then the group $f_*(\pi_1(V))$ contains a nilpotent subgroup of finite index. This follows from the above remark about the loops and basic geometry of $K < 0$ (see [B-G-S]).

Another (more important) case where (B) can be ruled out was pointed out by Corlette (see $[\mathrm{Cor}]_2$).

4.5.D_2. *Let* $W = Y/\Gamma$ *where* Y *is a symmetric space. If the Zariski closure of the subgroup* $f_*(\pi_1(V))$ *in the real algebraic group* Iso Y (*containing* $\pi_1(W) \supset f_*(\pi_1(V))$ *is semisimple then* (B) *is impossible and so* $f(v, 0)$ *is homotopic to a harmonic map. Furthermore this remains valid for the general* Γ*-invariant case where* X/Γ *is compact and* Γ *the Zariski closure of* Γ in Iso Y *is semisimple* (see $[\mathrm{Cor}]_1$).

Remark. The existence theorem for compact W is due to Eells and Sampson. Although the proof of the non-compact generalization uses the same analytic techniques, it provides us with a by far more powerful tool for the (super) rigidity problem.

4.6. Special Bochner formulas

4.6.A. We come back to the Riemannian geometric formulation of the superrigidity problem, alluded to in 2.9.A.

In this problem, representations of lattices into compact groups are neglected, as reflected in the following definition.

Definition : Let G, G' be semi simple Lie groups, Γ a subgroup of G, $\rho : \Gamma \to G'$ a homomorphism.

A virtual extension of ρ to G is the following data :

- an extension $1 \to K_1 \to G_1 \xrightarrow{\pi} G \to 1$ of G by a compact Lie group K_1
- an embedding $i : \Gamma \to G_1$ such that $\pi \circ i = \text{identity}$
- a homomorphism $h : G_1 \to G'$ such that

$$\rho = h \circ i .$$

Note that, since the group of outer automorphisms of K_1 is discrete, an extension G_1 as above is a direct product, up to finite groups.

Proposition : *Let* X, X' *be symmetric spaces with negative Ricci curvature. Let* G, G' *be their isometry groups,* $\Gamma \subset G$ *a subgroup,* $\rho : \Gamma \to G'$ *a homomorphism. Then* ρ *admits a virtual extension to* G *if and only if there exists a totally geodesic, Γ-equivariant map* $X \to X'$.

Proof.

1) *The virtual extension arising from a totally geodesic map.* Let $f : X \to X'$ be totally geodesic. If X is irreducible, then f is a homothety. Indeed, the pulled back metric

$$f^* g_{X'}$$

is parallel on X. If X is not irreducible, it splits as a riemannian product $X = X_1 \times X_2$, f factors through the projection $pr_1 : X \to X_1$ and a totally geodesic embedding $f' : X_1 \to X'$. The isometry group of $Y = f'(x_1)$ is a direct factor G_f in $G = G_f \times H$.

Let $G_2 \subset G'$ be the subgroup that fixes Y globally. Then G_2 acts by isometries on Y, which gives a compact extension

$$I \to K_1 \to G_2 \to G_f \to 1$$

then $G_1 = G_2 \times H$ is a compact extension of G with a homomorphism

$$h : G_1 \xrightarrow{pr_1} G_2 \subset G'.$$

If f is Γ-equivariant, then $\rho(\Gamma) \subset G_2$. Let $pr_2 : \Gamma \to G = G_f \times H \to H$ be the projection to the H factor then

$$i = \rho \times pr_2 = \Gamma \to G_2 \times H = G_1$$

embeds Γ into G_1, $\pi \circ i =$ identity, $\rho = h \circ i$ so that h is a virtual extension of ρ.

2) *The totally geodesic orbit of a virtual extension.* Let $h : G_1 \to G'$ be a virtual extension of ρ. Then

$$G_2 = h(G_1)$$

is a reductive subgroup of G'. We use a theorem of G.D. Mostow ([Mos]$_5$, theorem 6) : G_2 is left invariant by some Cartan involution θ of G'. The involution θ acts on X' as the geodesic symmetry through a point y. Let Y be the orbit of y under G_2, and II its second fundamental form. We see that

$$\theta^* II_y = - II_{\theta(y)}$$

and so $II_y = 0$, $II = 0$ everywhere by homogeneity, and Y is totally geodesic.

Let $\pi = G_1 \to G$ be the extension. Fix an origin x_0 in X, with stabilizer K. Then $h(\pi^{-1}K)$ is a compact subgroup of isometries of Y. Since Y is a symmetric space with $K \leq 0$, $h(\pi^{-1}K)$ fixes some point y_0 in Y. As a consequence, h descends to a homomorphism $\ell : G \to \mathrm{Isom}(Y)$ and an equivariant map $f = X \to Y$ can be defined by

$$f(gx_0) = \ell(g)y_0 \,.$$

The quadratic form f^*g_y on X is G-invariant thus parallel with respect to g_x, i.e., f is totally geodesic.

Finally we check that f is Γ-equivariant :

$$\text{for} \qquad g \in G, \gamma \in \Gamma\,, \ \phi(\gamma g x_0) = \ell(\gamma)\, \ell(g)\, y_0$$

$$= h(i(\gamma))\, f(x)$$

$$= \rho(\gamma)\, f(x)\,.$$

4.6.B. In view of 4.6.A. and 4.5, the superrigidity problem translates into the following terms :

Let f be a (twisted) harmonic map $V \to W$ where V is compact. Assume the curvature of W is sufficiently negative (e.g. W is a locally symmetric space of non compact type). For special V (e.g. cerain types of locally symmetric spaces), can one conclude that f satisfies some extra equations ? Ultimately, that f is totally geodesic ?

Let us introduce suggestive notations. Let $E = f^*\, TW$ denote the pull back of the tangent bundle of W. It is a vector bundle on V, equipped with a metric and a connection. We view

$$\alpha = df \in C^\infty(T^*V \otimes E)$$

as a vector valued 1-form on V. The connection on V and E determine a connection ∇ on $T^*V \otimes E$. Then the Hessian of f is $\nabla\alpha$, a vector valued bilinear form on V, compare 4.2.A_2'.

Bilinear forms on V splits as

$$T^*V \otimes T^*V = \Lambda^2 \oplus \mathbb{R} g_V \oplus S_0^2$$

into skew symmetric forms, the metric, and tracefree symmetric forms.

Accordingly, the Hessian $\nabla\alpha$ splits as

$$\nabla\alpha = d\alpha + d^*\alpha + (\nabla\alpha)^{S^2_0}$$

and

$$\alpha = df \text{ implies } d\alpha = 0$$

$$f \text{ is harmonic if and only if } d^*\alpha = 0$$

$$f \text{ is totally geodesic if and only if } \nabla\alpha = 0.$$

Why should $d\alpha = 0$ and $d^*\alpha = 0$ imply that the rest of ∇a vanishes ? This is where compactness of V and Bochner formulas play a role.

4.6.B_1. Bochner's vanishing theorem for Riemannian manifolds with nonnegative Ricci curvature.

We illustrate the method of Bochner formulas of an example. Let V be a compact Riemannian manifold, and α a harmonic 1-form on V. Then the Bochner formula states that , if $e = \frac{1}{2}\|\alpha\|^2$, then

$$\Delta e = \|\nabla\alpha\|^2 + \mathrm{Ricci}(\alpha, \alpha)$$

where the Ricci tensor is thought of as a quadratic form on T^*V , see [Boc-Yan].

If we assume that the Ricci curvature of V is nonnegative, we conclude that $\nabla\alpha = 0$ everywhere. Indeed, on a compact manifold, the Laplacian of a function integrates to zero.

The above argument can be generalized to include vector valued forms. The extra data is a vector bundle E over V equipped with a metric and a connection. For an E-valued 1-form α, the Bochner formula becomes

$$\Delta e = \|\nabla\alpha\|^2 + \mathrm{Ricci}(\alpha, \alpha) - \langle R^E, \alpha \wedge \alpha\rangle$$

where the extra term involves the curvature of E,

$$R^E \in \Lambda^2 T^*V \otimes \mathrm{End}\, E$$

contracted with $\alpha \wedge \alpha \in \Lambda^2 T^*V \otimes E \otimes E$.

For flat bundles, the formula is unchanged.

When f is a map $V \to W$ and $E = f^* TW$, $\alpha = df$, we recover the Eells-Sampson formula used in 4.5.D. Indeed, if $R^W \in \Lambda^2 T^*W \otimes \mathrm{End}\, TW$ is the curvature tensor of W, the curvature of E is

$$R^2 = f^* R^W = (\Lambda^2 \alpha)^* R^W$$

and

$$<R^E, \alpha \wedge \alpha> = Q_W$$

(in the notation of 4.5.D) is nonnegative provided W has nonpositive sectional curvature. We infer that every harmonic map from V to W is totally geodesic provided :

W has nonpositive sectional curvature ;

V has nonnegative Ricci curvature.

This theorem is of no use when V is locally symmetric of non compact type, since then $\mathrm{Ricci}_V < 0$. The point of the special formulas we explain next is that the sign of the curvature of V does not interfere. What is required instead is that V have special holonomy.

4.6.B_2 Kähler manifolds and pluriharmonic maps.

A Kähler manifold of real dimension 2n has holonomy a subgroup of U(n). Under U(n), trace free symmetric bilinear forms split as

$$S^2_0 = S^{2,0+0,2} \oplus S^{1,1}_0$$

where the summands are the eigenspaces of the involution

$$b(u, v) \mapsto b(Ju, Jv)$$

(J denotes the complex structure).

Thus, on a Kähler manifold, there is a new equation, i.e., a new linear condition that one can impose on the second derivatives of a map.

4.6.B_3. Definition : Let V be a Kähler manifold, W a Riemannian manifold. Say a map $f : V \to W$ is *pluriharmonic* if its Hessian $\nabla\alpha$ is of type $S^{2,0+0,2}$, i.e., its components on Λ^2 and $S^{1,1}$ vanish.

When V has real dimension 2, pluriharmonic means harmonic. In higher dimensions, f is pluriharmonic if and only if its restriction to every (germ of) holomorphic curve in V is harmonic. As a consequence, pluriharmonicity does not depend on the particular Kähler metric on V, only on the complex structure of V. In fact, a real valued function on V is pluriharmonic if and only if it is locally the real part of a holomorphic function. Thus the familiar Hodge therorem

$$H^1(V, \mathbb{C}) = H^{1,0} \oplus H^{0,1}$$

says that, on a compact Kähler manifold, every harmonic 1-form is pluriharmonic.

The main step in Y.T.Siu's 1980 theorem is an extension to vector valued 1-forms of this property.

4.6.C. Negative curvature operator.

We describe now the kind of negativity required on W.

Definition : The curvature tensor of a Riemannian manifold can be viewed as a quadratic form on exterior 2-forms. It is then called the *curvature operator* $\check{R}$.

Example : Symmetric spaces of non compact type have $\check{R} \le 0$.

Proof : Let $X = G/K$ be symmetric. Let $\underline{p}$ be the orthogonal complement of $\underline{k}$ in $\underline{g}$ with respect to the Killing form B. B is positive definite on $\underline{p}$, it gives rise to the Riemannian metric on X. Consider the Lie bracket as a map

$$[\ ,\ ] = \Lambda^2 \underline{p} \to \underline{k}\,.$$

then the curvature operator of X - a quadratic form on $\Lambda^2 \underline{p}$ - is the pull back of $B_{|\underline{k}}$ by $[\ ,\ ]$. As a consequence, it is non positive, and its kernel coincide with kernel of $[\ ,\ ]$.

4.6.D. A vanishing theorem for harmonic maps of manifolds with special holonomy

In 1980, Y.T. Siu found a generalization of the Hodge theorem $H^1(V, \mathbb{C}) = H^{1,0} \oplus H^{0,1}$ for harmonic maps of Kähler manifolds to hermitian locally symmetric spaces, and gave striking consequences, including a new proof of Mostow rigidity for hermitian locally symmetric spaces (see below 4.6.F_2). Later, J. Sampson observed that Siu's method could be extended to the case where the range is not Kähler. Recently, K. Corlette has included Siu's theorem into the wider framework of harmonic maps from manifolds with special holonomy.

4.6.D_1. Theorem. [Siu], [Sam], $[\mathrm{Cor}]_1$. *Let* V *be a compact Riemannian manifold that admits a parallel form* ω. *Let* W *be a Riemannian manifold with non positive curvature operator. Then every (twisted) harmonic map* $f : V \to W$ *satisfies*

$$d^*(df \wedge \omega) = 0\,.$$

This equation in a generalisation of the notion of pluriharmonicity. Indeed, let us compute $d^*(df \wedge \omega)$. This is a linear expression in $\nabla\alpha$, $\alpha = df$, and we can assume α is scalar, V is flat. Let e_i be an orthonormal frame

$$\begin{aligned} d^*(\alpha \wedge \omega) &= -\sum_i e_i^* \;\; \nabla_{e_i}(\alpha \wedge \omega) \\ &= -\sum_i e_i^* \;\; \left(\nabla_{e_i}\alpha\right) \wedge \omega \\ &= \left(-\sum_i e_i^* \;\; \nabla_{e_i}\alpha\right) \wedge \omega - \sum_i \left(\nabla_{e_i}\alpha\right) \wedge \left(e_i^* \;\; \omega\right) \\ &= \left(d^*\alpha\right)\omega - \mathrm{ad}_{\nabla\alpha}(\omega)\,; \end{aligned}$$

where we view $\nabla\alpha$ as an endomorphism of TV and the natural action of $\nabla\alpha$ on TV is extended as a derivation of the exterior algebra.

When V is a Kähler manifold and ω its Kähler form, we identify 2-forms with endomorphisms via the metric. Then ω becomes the complex structure J, and $\mathrm{ad}_{\nabla\alpha}(\omega)$ becomes

$$\nabla\alpha \circ J - J \circ \nabla\alpha, \text{ i.e. } \mathrm{ad}_{\nabla\alpha}(\omega) = (\nabla\alpha)^{1,1}.$$

Thus, for a harmonic 1-form α,

$$d^*(\alpha \wedge \omega) = 0 \Leftrightarrow \nabla\alpha \in S^{2,0+0,2}$$

$$\Leftrightarrow \alpha \text{ is pluriharmonic}.$$

Proof of vanishing theorem 4.6.D_1

The idea is that exterior multiplication with a parallel form commutes with the Laplacian, see[Che]. It is convenient to use first Clifford multiplication instead of exterior multiplication.

4.6.D_3. The Clifford Formalism (see [Law-Mic]).

Clifford multiplication is an associative algebra structure on $\Lambda^* T^* V$, the space of exterior forms on a Euclidean vector space. Recall that interior multiplication of exterior forms α and β is defined by the following identity for all exterior forms γ,

$$<\alpha \quad \beta, \gamma> = <\beta, \alpha \wedge \gamma>;$$

i.e., interior multiplication by α is the adjoint of exterior multiplication by α.

For a 1-form α and p-form ω, one defines the Clifford product

$$\alpha.\omega = \alpha \wedge \omega + \alpha \quad \omega. \qquad (*)$$

This extends to an associative multiplication on $\Lambda^* T^* V$.

The *Dirac operator* $\mathcal{D}$, a first order operator on differential forms, is the composition of the covariant derivative : $C^\infty(\Lambda^* T^* V) \to C^\infty(T^* V \otimes \Lambda^* T^* V)$ with Clifford multiplication :

$$T^* V \otimes \Lambda^* T^* V \to \Lambda^* T^* V$$

In other words, given an orthonormal basis e_i of TV, with dual basis e_i^*,

$$\mathcal{D}\beta = \sum_i e_i^* . \nabla_{e_i} \beta$$

Formula (*) shows that

$$\mathcal{D} = d + d^*$$

Let ω be a parallel p-form on V.

Denote by C the operator on differential forms defined by right Clifford multiplication with ω. Clearly

$$\mathcal{D} C\beta = \sum_i e_i^* \,.\, \nabla_{e_i}(\beta.\omega) = C\mathcal{D}\beta$$

as follows form the associativity of Clifford multiplication.

Thus the Laplacian $\Delta = (d + d^*)^2 = \mathcal{D}^2$ commutes with C. Splitting

$$C = C^p + C^{p-2} + C^{p-4} + \dots ,$$

according to the degree of forms, we see that C^p, exterior multiplication with ω, commutes with the Laplacian : $[\Delta, C^p] = 0$.

Some of the more basic identities obtained when splitting

$$[\mathcal{D}, C] = 0$$

according to the degree of forms,

$$[d, C^p] = 0 ;$$

$$[d, C^{p-2}] + [d^*, C^p] = 0 ;$$

and so on, are classical.

The first identity, $d(\alpha \wedge \omega) = (d\alpha) \wedge \omega$, is very general. In the Kähler case, where $C^p = C^2$ is usually denoted by L, the second identity is the familiar

$$[d^*, L] = d^c.$$

All the above formulas extend without change to vector valued forms α (and scalar ω).

4.6.D_4. <u>The vanishing theorem, scalar case :</u>

On a compact manifold V, let ω be a parallel form and α a harmonic 1-form. Using identity

$$[d + d^*, C] = 0$$

we integrate by parts :

$$\int |dC\alpha|^2 = -\int \langle dC\alpha, d^*C\alpha \rangle$$

$$= -\int \langle d^2C\alpha, C\alpha \rangle$$

$$= 0$$

and thus $dC\alpha = d^*C\alpha = 0$. In particular,

$$d^*C^p\alpha = d^*(\alpha \wedge \omega) = 0$$

as announced.

4.6.D_5. The vanishing theorem, general case :

The point is to compute the sign of $<d^2C\alpha, C\alpha>$ when $\alpha = df$, f a map $V \to W$. Let R^W be the curvature tensor of W. It defines a quadratic form on $TW \otimes TW$ (which is non zero only on $\Lambda^2 TW$), which combined with the metric on TV gives rise to a quadratic form Q on

$$\Lambda^* T^* V \otimes TW \otimes TW$$

Claim :

$$<d^2C\alpha, C\alpha> = 2Q(\alpha.\alpha.\omega)$$

Clearly, this proves the theorem, since Q has the same sign as $\check{R}^W$.

This an algebraic computation. Splitting into degrees of forms, we find

$$<d^2C\alpha, C\alpha> = <d^2C^{p-2}\alpha, C^p\alpha> = <d^2(\alpha \quad \omega), \alpha \wedge \omega>$$

Now d^2 is exterior multiplication with the curvature of the bundle f^*TW, that is $R^W \circ (\alpha \wedge \alpha)$. We get

$$<d^2C\alpha, C\alpha> = <R^W((\alpha \wedge \alpha) \wedge (\alpha \quad \omega)), \alpha \wedge \omega> .$$

Using symmetries of R^W, the Bianchi identity $d^2\alpha = 0$, and symmetries of the expression $<\alpha \wedge \alpha \wedge (\alpha \quad \omega), \alpha \wedge \omega>$, this becomes

$$2 < R^W(\alpha \wedge \quad (\alpha \quad \omega), \alpha \quad (\alpha \wedge \omega)>$$

$$= 2Q(\alpha.\alpha.\omega) .$$

4.6.E. Negative complex curvature

In the Kähler case, i.e. V is Kähler, ω its Kähler form, the curvature assumption on W in theorem 4.6.D_1 can be weakened. Indeed, $\alpha \quad \omega = J\alpha$ and we find that the quadratic form Q is the pull back of the quadratic form $\check{R}^W$ on $\Lambda^2 TW$ by the map

$$\alpha \wedge J\alpha = \Lambda^2 TV \to \Lambda^2 TW$$

Given a unitary basis e_j of TV, denote by

$$Z_j = \alpha(e_j) + i\alpha(Je_j) \in TW \otimes \mathbb{C}$$

then

$$2Q(\alpha.\alpha.\omega) = 2Q(\alpha \wedge J\alpha)$$

$$= \sum_{i<j} < \check{R}^W(Z_i \wedge Z_j), \bar{Z}_i \wedge \bar{Z}_j >$$

thus for the previous argument to apply, it is sufficient to assume that

$$< \check{R}^W(Z_1 \wedge Z_2, \bar{Z}_1 \wedge \bar{Z}_2 > \leq 0$$

for all *complex* tangent vectors $Z_1, Z_2 \in TW \otimes \mathbb{C}$. When this condition is satisfied, one says that W has *nonpositive complex curvature.*

Under this assumptions (V compact Kähler, W has nonpositive complex curvature), theorem 4.6.D_1 is due to Sampson ([Sam]), elaborating on Siu's work where W was assumed Kähler as well.

4.6.F. Non linear equations satisfied by harmonic maps.

The proof of Theorem 4.6.D_1 relies on the identity

$$\int_V |dC\alpha|^2 - 2Q(\alpha.\alpha.\omega) = 0$$

where Q is non positive. Thus with the linear equations $dC\alpha = 0$ comes the nonlinear condition

$$Q(\alpha.\alpha.\omega) = 0\,.$$

For a Kähler domain V, the condition states that for all $Z_1, Z_2 \in df(T^{1,0}V)$, $Z_1 \wedge Z_2 \in \operatorname{Ker} \check{R}^W \otimes \mathbb{C}$, and has the following interpretation : the Levi Civita connection endows the bundle f^*TW with the structure of a holomorphic bundle. Furthermore the (1, 0)-component of df is a holomorphic 1-form with values in f^*TW (on other formulation of the fact that f is pluriharmonic). When the range W is a symmetric space G/K, the nonlinear condition says that the subspace $df(T^{1,0}V) \subset \underline{p} \otimes \mathbb{C}$ is an abelian subalgebra (compare 4.6.C). This has been analyzed by Siu, Carlson-Toledo.

4.6.F_2 Proof of Mostow rigidity for hermitian symmetric spaces, following Siu.

Let V, W be compact hermitian locally symmetric spaces. Let $f_0 : V \to W$ be a homotopy equivalence. Deform f_0 to a harmonic map f. Since the degree of f is non zero, the differential df is surjective at many points.

Lemma (Carlson-Toledo [Car-Tol1]). *The only abelian subspaces of* $\underline{p}^W \otimes \mathbb{C}$ *of maximal dimension are* $\underline{p}^{1,0}$ *and* $\underline{p}^{0,1}$. *i.e., the subspaces that define the complex structure on* W.

Admitting this fact, we see that f is holomorphic or antiholomorphic on a big set. By elliptic regularity, f is holomorphic (or antihomorphic) everywhere. f is injective since its fibers are complex subvarieties homologous to a point. Lift f to a biholomorphism of symmetric spaces, which are bounded domains in $\mathbb{C}^n$. The symmetric metric coincides with the Bergman metric, which is natural under biholomorphisms. We conclude that f is an isometry.

The assumption that V is locally symmetric was needed to conclude that f_0 was homotopic to an isometry only. For a Kähler domain V, one concludes that f_0 is homotopic to a holomorphic or antiholomorphic map.

4.6.F_3. Non linear Hodge theory.

Hodge theory is concerned with the interplay between topological objects (de Rham cohomology) and holomorphic objects (cohomology of sheaves of holomorphic forms). Nonlinear Hodge theory extends this correspondance and gives holomorphic counterparts to space of representations of fundamental groups of compact Kähler manifolds.

Let V be compact Kähler. Start with a finite dimensional representation

$$\rho : \pi_1(V) \to Gl(r, \mathbb{C}).$$

Assume the Zariski closure of $\rho(\pi_1(V))$ in $GL(r, \mathbb{C})$ is reductive. The representation ρ determines a vector bundle

$$F = \tilde{V} \times_\rho \mathbb{C}^r$$

over V. Hermitian metrics on F correspond to twisted maps of V to the space X of hermitian quadratic forms on $\mathbb{C}^r$, i.e., up to an irrelevant scale, the symmetric space $Gl(r, \mathbb{C})/U(r)$.

According to theorem 4.5.D_2, there exists on F a unique "harmonic metric" (i.e., metric corresponding to a harmonic equivariant map f from $\tilde{V}$ to X). The bundle F carries a natural flat connection ∇ (since it lifts to a trivial bundle on $\tilde{V}$). In general, this connection is not unitary with respect to the harmonic metric, thus it splits as

$$\nabla = D + \theta$$

where D is a unitary connection, θ a 1-form with values in (symmetric) endomorphism of F, which is nothing but df.

In this context, paragraph 4.6.F_1 can be improved : D defines a holomorphic structure on F (we recover the general fact that E = End F has a holomorphic structure), and $\theta^{1,0}$ is a End F-valued holomorphic 1-form. Furthermore,

$$(\theta^{1,0})^2 = 0.$$

Such data $(E, \theta^{1,0})$ form a Higgs bundle ([Hit], [Sim]).

When the representation ρ is irreducible, the Higgs bundle is stable.

Conversely, a stable Higgs bundle whose Chern classes vanish arises from a representation of $\pi_1(V)$ (ibidem).

4.6.G. Rigid exterior forms and superrigidity

4.6.G$_1$. Definition : Let ω be an exterior form on the vector space $\mathbb{R}^n$. We say that ω is *rigid* if the subgroup of $Gl(n, \mathbb{R})$ that fixes ω is compact.

Example : The Kähler form on $\mathbb{C}^n$ is not rigid. Indeed, its stabilizer is the non compact symplectic group $Sp(2n, \mathbb{R})$.

Example : On quaternionic vector space $\mathbb{H}^n$, there is a $Sp(n-1)\,Sp(1)$-invariant 4-form, which is rigid if $n \geq 2$. Let $\omega_i, \omega_j, \omega_k$ denote the three components of the $\mathrm{Im}\mathbb{H}$-valued 2-form

$$\mathrm{Im}\left(\sum_{\ell} \overline{q_\ell} q'_\ell\right)$$

then

$$\omega = \omega_i^2 + \omega_j^2 + \omega_k^2$$

has stabilizer exactly $Sp(n-1)\,Sp(1)$ provided $n \geq 2$.

Similarly, there is on $\mathbb{C}a^2$ a rigid 8-form, whose stabilizer is exactly $Spin(9) \subset SO(16)$.

A Riemannian manifold of dimension $4n$ is called quaternion-Kähler if its holonomy at each point is contained in a copy of $Sp(n-1)\,Sp(1) \subset Gl(4n, \mathbb{R})$. Such a manifold automatically carries a rigid parallel 4-form. Quaternion-Kähler symmetric spaces are listed in ([Bes], chap. 14). These include quaternionic hyperbolic spaces $\mathbb{H}^{4n}$, $n \geq 2$. It is unknown whether there exist non locally symmetric compact quaternionic manifolds with negative Ricci curvature.

4.6.G$_2$. Theorem : *Let* V *be a compact Riemannian manifold that admits a rigid paralled form. Let* W *have non positive curvature operator. Then every twisted harmonic map of* V *to* W *is totally geodesic.*

Proof : Theorem 4.6.D$_1$ applies to give $d^*(df \wedge \omega) = 0$. Formula 4.6.D$_2$ implies that the Hessian $\nabla df \in A \otimes f^*TW$ where $A \subset \mathrm{End}\, TV$ is the Lie algebra of the stabilizer of ω in TV. If A integrates to a compact group, then every matrix in A has imaginary eigenvalues. But $ddf = 0$ implies ∇df is symmetric and has real eigenvalues. We conclude that $\nabla df = 0$, i.e. f is totally geodesic.

Corollary; (Archimedean superrigidity for quaternionic and Cayley hyperbolic spaces).
Let G *be equal, either to* $Sp(n, 1)$, $n \geq 2$ or F_4^{-20}. *Let* $\Gamma \subset G$ *be a discrete cocompact subgroup. Let* G' *be a semisimple Lie group with maximal compact subgroup* K'.
Let ρ *be a homomorphism of* Γ *into* Γ'. *Assume that the Zariski closure of* $\rho(\Gamma)$ *in* G' *is reductive. Then* ρ *factors through a homomorphism*

$$\Gamma \to G \times K' \to G'$$

Remark : (a) The corollary extends to finite covolume lattice, see ($[Cor]_1$).

(b) Many symmetric spaces of higher rank admit rigid parallel forms, but not all of them.

References

[Ale] A.D. Aleksandrov, Convex Polytopes, Moscow 1950 (Russian).

[And] E. Andreev, Convex polyhedra of finite volume in Lobachevski space (Russian), Mat. Sb. 83 (125) (1970) p.p. 256-260.

[Bas] H. Bass, The congruence subgroup problem, Proc. Conf. Local Fields, Driebergen 1966, Springer 1967, p.p. 16-22.

[B-G-S] W. Ballmann, M. Gromov, V. Schroeder, Manifolds of non-positive curvature. Progress in Math. Vol. 61, Birkhäuser, Boston-Basel-Stuttgart (1985).

[Bes] A.L. Besse, Einstein manifolds, Ergeb. Bd. 10, Springer Verlag, 1986.

$[\mathrm{Bor}]_1$ A. Borel, Introduction aux groupes arithmétiques, Hermann, Paris 1969.

$[\mathrm{Bor}]_2$ A. Borel, Compact Clifford-Klein forms of symmetric spaces, Topology 2, (1963), p.p. 111-122.

[Bor-Sha] Z.I. Borevič and I.R. Shafarevič, Number theory, Acad. Press 1966.

[Bus] H. Buseman, Extremals on closed hyperbolic space forms, Tensor, 16 (1965), p.p. 313-318.

[B-Y] S. Bochner and K. Jano, Curvature and Betti numbers, Princeton University Press, Princeton 1953.

[Cal] E. Calabi, On compact Riemannian manifolds with constant curvature I. Proc. Symp. Pure Math. Vol III, p.p. 155-180, A.M.S. Providence, R.I. 1961.

[Cal-Ves] E. Calabi and E. Vesentini, On compact locally symmetric Kähler manifolds, Ann. of Math. (2) 71 (1960), p.p. 472-507.

[Car-Tol] J. Carlson, D. Toledo, Harmonic maps of Kähler manifolds to locally symmetric spaces, Publ. Math. I.H.E.S. 69 (1989), p.p. 173-201.

[Cas] J.W.S. Cassels, An introduction to the geometry of numbers, Springer-Verlag 1959.

[Ch-Eb] F. Cheeger, D. Ebin, Comparison theorems in Riemannian geometry, North-Holland 1975.

[Che] S.S. Chern, On a generalization of Kähler geometry, in "Algebraic Geometry and Topology Symp. in honor of S. Lefschetz", Princeton Univ. Press, (1957), p.p. 103-121.

$[\mathrm{Cor}]_1$ K. Corlette, Archimedean superrigidity and hyperbolic geometry, preprint 1990.

$[\mathrm{Cor}]_2$ K. Corlette, Flat G-bundles and canonical metrics, J. Diff. Geom. 28 (1988), p.p. 361-382.

[dlH-V] P. de la Harpe et A. Valette. La propriété (T) de Kazhdan pour les groupes localement compacts, Astérisque 175 (1989). Soc. Math. de France.

[Don] S. Donaldson, Twisted harmonic maps and the self duality equations, Proc. London Math. Soc. 55 (1987), p.p. 127-131.

$[\mathrm{E\text{-}L}]_1$ J. Eells and L. Lemaire, A report on harmonic maps, Bull. Lond. Math. Soc. 10 (1978), p.p. 1-68.

[E-L]$_2$ J. Eells and L.Lemaire, Another report on harmonic maps, Bull. Lond. Math. Soc. 20 (1988), p.p. 385-525.

[E-S] J. Eells and J.H. Sampson, Harmonic mappings of Riemannian manifolds, Amer. J. Math. 86 (1964), pp. 109-160.

[Efz] V. Efremovič, The proximity geometry of Riemannian manifolds, Uspehi Mat., Nauk 8 (1953), p. 189 (Russian).

[Ef-Ti] V. Efremovič and E. Tichomirova, Equimorphisms of hyperbolic spaces, Isv. Ac. Nauk. 28 (1964), p.p. 1139-1144.

[Fl] W. J. Floyd, Group completion and limit sets of Kleinian groups, Inv. Math. 57 (1980), p.p. 205-218.

[Für]$_1$ H. Fürstenberg, Poisson boundaries and envelopes of discrete groups, Bull. Am. Math. Soc. 73 (1967), p.p. 350-356.

[Für]$_2$ H. Fürstenberg, Rigidity and cocycles for ergodic actions of semisimple Lie groups. Lect. Notes Math. 842, p.p. 273-292, Springer-Verlag 1981.

[Für]$_3$ H. Fürstenberg, Boundary theory and stochastic processes on homogeneous spaces. Proc. Symp. Pure Math. XXVI, 1973, p.p. 193-233.

[Geh] F.W. Gehring, Rings and quasiconformal mappings in space, Trans. Amer. Math. Soc. 103 (1962), p.p. 353-393.

[Gro]$_1$ M. Gromov, Hyperbolic manifolds according to Thurston and Jorgensen, in Springer Lecture Notes, 842 (1981), p.p. 40-53.

[Gro]$_2$ M. Gromov, Groups of polynomial growth and expanding maps, Publ. Math. IHES, 53 (1981), p.p. 53-78.

[Gro]$_3$ M. Gromov, Hyperbolic manifolds groups and actions, in Riem. surfaces and related topics, Ann. Math. Studies 97 (1981), p.p. 183-215.

[Gro]$_4$ M. Gromov, Hyperbolic groups, in Essays in group theory, S.M. Gersten ed., p.p. 75-265, Springer-Verlag 1987.

[G-L-P] M. Gromov, J. Lafontaine, and P. Pansu, Structures métriques pour les variétés riemanniennes, CEDIC- Fernand Nathan, Paris,1981.

[G-P] M. Gromov and I. Piatetski-Shapiro, Non-arithmetic groups in Lobachevsky spaces, Publ. Math. I.H.E.S. 66 (1988), p.p. 93-103.

[G-S] M. Gromov, R. Schoen, in preparation.

[Ha-Mu] U. Haagerup and M. Munkholm, Simplices of maximal volume in hyperbolic n-space. Act. Math. 1941 (1981), p.p. 1-11.

[Hit] N. Hitchin, The selfduality equations on a Riemann surface, Proc. London Math. Soc. 55 (1987), p.p. 59-126.

[Kar] H. Karcher, Riemannian center of mass and mollifier smoothing, Comm. Pure and Appl. Math. 30 (1977), p.p. 509-541.

[Klin] W. Klingenberg, Geodätischer Fluss auf Mannigfaltigkeiten vom hyperbolischen Typ, Inv. Math. 14 (1971), p.p. 63-82.

[Law-Mic] B. Lawson and M.-L. Michelsohn, Spin geometry, Princeton University Press, Princeton 1989.

[Kaz] D. Kazhdan, Connection of the dual space of a group with the structure of its closed subgroups, Func. Anal. and Appl. 1 (1967), p.p. 63-65.

[Mal] A.I. Malcev, On a class of homogeneous spaces, Amer. Math. Soc. Transl. 39, (1951), p.p. 276-307.

$[\mathrm{Mar}]_1$ G. Margulis, Discrete groups of motions of manifolds of non-positive curvature (Russian), ICM 1974. Translated in A.M.S. Transl. (2) 109 (1977), p.p. 33-45.

$[\mathrm{Mar}]_2$ G. Margulis, Discrete subgroups of semisimple Lie groups, Springer-Verlag 1990.

$[\mathrm{Mar}]_3$ G. Margulis, The isometry of closed manifolds of constant negative curvature with the same fundamental group, Dokl. Akad. Nauk. SSSR 192 (1970), p.p. 736-737.

[Mau] F.I. Mautner, Geodesic flows on symmetric Riemannian spaces, Ann. of Math. 65 (1957), p.p. 416-431.

$[\mathrm{Mil}]_1$ J. Milnor, Whitehead torsion, Bull. Amer. Math. Soc. 72 (1966), p.p. 358-426.

$[\mathrm{Mil}]_2$ J. Milnor, A note on curvature and fundamental group, J. Diff. Geom. 2 (1968), p.p. 1-7.

[Mok] N. Mok, Metric rigidity theorems on Hermitian locally symmetric manifolds, World Sci. Singapore-New Jersey-London-Hong Kong, 1989.

[Mors] M. Morse, Geodesics on negatively curved surfaces, Trans. Amer. Math. Soc 22 (1921) p.p. 84-100.

$[\mathrm{Mos}]_1$ G. Mostow, Quasi-conformal mappings in n-space and the rigidity of hyperbolic space form, Publ. Math. I.H.E.S., vol. 34, 1968, p.p. 53-104.

$[\mathrm{Mos}]_2$ G. Mostow, Equivariant embeddings in Euclidean space, Ann. of Math. 65 (1957), 432-446.

$[\mathrm{Mos}]_3$ G. Mostow, Factor spaces of solvable groups, Ann. of Math. 60 (1954), p.p. 1-27.

$[\mathrm{Mos}]_4$ G.D. Mostow, Strong rigidity of symmetric spaces, Ann. Math. Studies 78, Princeton 1973.

$[\mathrm{Mos}]_5$ G.D. Mostow, Some new decomposition theorems for semisimple groups, Memoirs Amer. Math. Soc. 14 (1955), p.p. 31-54.

[Pal] R. Palais, Equivalence of nearby differentiable actions of a compact group, Bull. Am. Math. Soc. 61(1961), p.p. 362-364.

$[\mathrm{Pan}]_1$ P. Pansu, Croissance des boules et des géodésiques fermées dans les nilvariétés, Ergod. Th. Dynam. Syst. 3 (1983), p.p. 415-445.

$[\mathrm{Pan}]_2$ P. Pansu, Métriques de Carnot-Caratheodory et quasi-isométries des espaces symétriques de rang un, Ann. of Math. 129:1 (1989), p.p. 1-61.

[Pont] L.S. Pontryagin, Topological groups, Gordon and Breach 1966.

[Pra] G. Prasad, Strong rigidity of Q-rank 1 lattices, Inv. Math. 21 (1973), p.p. 255-286.

[Rag] M.S. Raghunathan, Discrete subgroups of Lie groups, Springer-Verlag, 1972.

[Resh] Yu. G. Reshetniak, On conformal mappings of a space, Dokl. Acad. Nauk. SSSR 130 (1960) p.p. 981-983 = Soviet. Math. Dokl. 1 (1960), p.p. 122-124.

[Rei] H.M. Reimann, Capacities in the Heisenberg group, Preprint University of Bern (1987).

[Sel] A. Selberg, On discontinuous groups in higher dimensional symmetric spaces, Contribution to Function Theory, Tata Inst., Bombay 1960, p.p. 147-164.

[Sha] I. R. Shafarevič, Basic Algebraic geometry, Springer-Verlag 1974.

[Sam] J.H.Sampson, Applications of harmonic maps to Kähler geometry, Contemp. Math. 49 (1986), p.p. 125-133.

[Sim] C. Simpson, Constructing variations of Hodge structure using Yang-Mills theory and applications to uniformization, J. of the Amer. Math. Soc. 1 (1988), p.p. 867-918.

[Siu] Y.-T. Siu, The complex analyticity of harmonic maps and the strong rigidity of compact Kähler manifolds, Ann. of Math. 112 (1980), p.p. 73-111.

[Str] R. Strichartz, Sub-Riemannian geometry, Journ. of Diff. Geom. 24, (1986), p.p. 221-263.

[Šv] A. Švarc, A volume invariant of coverings, Dokl. Akad. Nauk SSSR 105 (1955), p.p. 32-34.

[Th] W. Thurston, Geometry and topology of 3-manifolds, Lecture notes Princeton, 1978.

[Tuk] P. Tukia, Quasiconformal extension of quasisymmetric maps compatible with a Möbius group, Acta Math. 154 (1985), p.p. 153-193.

[Tuk-Väi] P. Tukia, J. Väisälä, A remark on 1-quasiconformal maps, Ann. Acad. Sci. Fenn. 10 (1985), p.p. 561-562.

[Ul] S.M. Ulam, A collection of mathematical problems. Interscience Publ. N.Y. 1960.

[Väi] J. Väisälä, Lectures on n-dimensional quasiconformal mappings, Lecture Notes in Math. Vol 129, Springer, Berlin (1971).

[Wang] H.-C. Wang, Topics on totally discontinuous groups, in "Symmetric spaces" edited by W.M. Boothby and G.L. Weiss, p.p. 459-487, Marcel Dekker, Inc. N.Y. 1972.

[Wei] H. Weil, On discrete subgroups of Lie groups II, Ann. of Math. vol. 75 (1962), p.p. 578-602.

Instanton Invariants and Algebraic Surfaces

Christian Okonek

Mathematisches Institut
der Universität Bonn
Wegelerstraße 10
5300 Bonn
F.R.G.

Table of Contents

1 Introduction

2 Instantons
- 2.1 Yang-Mills connections
- 2.2 Moduli spaces of instantons

3 Donaldson invariants
- 3.1 Definition of the polynomials
- 3.2 Dependence on the metric

4 Hermitian-Einstein structures and stable bundles
- 4.1 Hermitian-Einstein bundles
- 4.2 Stable algebraic bundles
- 4.3 Instantons over algebraic surfaces

5 C^∞-structures of algebraic surfaces
- 5.1 Topology of algebraic surfaces
- 5.2 Exotic C^∞-structures on $\mathbb{P}^2\#9\bar{\mathbb{P}}^2$
- 5.3 Indecomposability of algebraic surfaces
- 5.4 Diffeomorphism groups

6 Floer homology
- 6.1 Instanton holomogy
- 6.2 Seifert fibered homology spheres

7 Donaldson-Floer polynomials and singularities
- 7.1 Relative invariants
- 7.2 Surface singularities

8 References

1 Introduction

These lecture notes have been written for a course of the C.I.M.E. session "Recent developments in geometric topology and related topics", June 4 - 12, 1990 in Montecatini Terme.
My intention was to give an introduction to recent developments in 3- and 4-dimensional geometry for an audience with some general background in geometry or topology.
These developments are all based on S.K. Donaldson's fundamental idea to use instanton moduli spaces as a tool for studying the geometry of low-dimensional manifolds.
His ideas have led to particular strong results in the integrable case, i.e. for algebraic surfaces: Instantons over algebraic surfaces can be interpreted as stable algebraic vector bundles. This correspondence, which ties the geometry of an algebraic surface to the differential topology of its underlying 4-manifold, allows to investigate instanton moduli spaces with algebraic techniques.
These things will be explained in the first three sections of these notes.
Applications to the differential topology of algebraic surfaces will be discussed in the next paragraph. The main results are the existence of infinitely many exotic C^∞-structures on some topological 4-manifolds, Donaldson's indecomposability theorem for algebraic surfaces, and computations of diffeomorphism groups of certain complete intersections.
The final two sections are concerned with the decomposition of 4-manifolds along homology 3-spheres. A. Floer has defined a new homology theory for such 3-manifolds using gauge theoretic constructions in dimension 3 and 4. His instanton homology can be computed explicitly for Seifert fibered homology spheres. Forthcoming work of Donaldson will define new invariants for certain 4-manifolds with homology sphere boundaries.
These invariants, which are homogeneous polynomials on the second homology of the 4-manifold with values in the Floer homology of the boundary, have been announced in an expository article by M. Atiyah; details are still unpublished. After a brief description of the Donaldson-Floer polynomials, which is based on Atiyah's article, the main results of a joint paper with W. Ebeling will be sketched. In this paper we study the relative polynomial invariants for Milnor fibers of 2-dimensional isolated complete intersection singularities and for the neighborhood of the curve at infinity in the weighted homogeneous case.
I like to emphasize that the basic results which I describe in these notes are due to Donaldson [D1], [D2], [D3], [D4], [D5]. In addition to his papers I have mainly used articles by Atiyah [A], Floer [F], Fintushel-Stern [FS2], [FS3], Friedman-Morgan [FM2], [FM4], Taubes [T], Uhlenbeck [U1], [U2], Ebeling-Okonek [EO1], [EO2], Okonek-Van de Ven [OV1], [OV2], [OV3], and Okonek [O].
The material in the first sections overlaps to some extent with the survey article [OV3].
I have tried to present the results in a non-technical way, ignoring in particular all analytical difficulties, like e.g. Sobolev completions. For these points, as well as for details of most of the proofs I have to refer to the original papers. Most

of the necessary background material can be found in the books Barth-Peters-Van de Ven [BPV], Freed-Uhlenbeck [FU] Hirzebruch [Hi], Kobayashi [K2] and Okonek-Schneider-Spindler [OSS].
I like to thank my wife Christiane for her help with the processing of these lecture notes.

2 Instantons

In this section we introduce some terminology, define Yang-Mills connections and instantons, and explain the basic properties of their moduli spaces.

2.1 Yang-Mills connections

Let X be a closed connected oriented differentiable 4-manifold, G a compact Lie group. For simplicity we will assume X is simply-connected and G is one of the groups S^1, $SU(2)$, $U(2)$, or $PU(2)$.
Consider a principal bundle P over X with structure group G. The **adjoint bundle** is the bundle $ad(P) = P \times_{\mathrm{Ad}} LG$ associated to P via the adjoint representation of G on its Lie algebra LG.
We denote by $\mathcal{A}(P)$ the set of all G-**connections** on P; this is an infinite dimensional affine space $\mathcal{A}(P) = A_0 + A^1(ad(P))$ modelled after the vector space $A^1(ad(P))$ of 1-forms with values in $ad(P)$ [FU].
A connection $A \in \mathcal{A}(P)$ induces **differential operators**

$$D_A : A^p(E) \to A^{p+1}(E)$$

satisfying the Leibniz rule

$$D_A(\varphi \cdot s) = d\varphi \otimes s + (-1)^p \varphi \wedge D_A(s)$$

on all vector bundles $E = P \times_\rho V$ associated to P.
The **curvature** of A is the 2-form $F_A \in A^2(ad(P))$ representing the composition $D_A \circ D_A$; it satisfies the **Bianchi identity** $D_A(F_A) = 0$ [K2].
Let $\mathrm{Aut}\,(P)$ be the **gauge group** of P/X, i.e. the group of bundle automorphisms of P covering the identity on X. The gauge group acts on $\mathcal{A}(P)$ in a natural way, so that the differential operator $D_{A\cdot f}$ of a transformed connection is given by

$$D_{A\cdot f} = f^{-1} \circ D_A \circ f.$$

The **isotropy group** $\mathrm{Aut}\,(P)_A$ of this action at a connection $A \in \mathcal{A}(P)$ is the centralizer of the holonomy of A and therefore contains the center Z_G of G [FU].
A connection is **irreducible** if its holonomy is dense in G; in this case one has $\mathrm{Aut}\,(P)_A = Z_G$. The group $\mathrm{Aut}^*(P) := \mathrm{Aut}\,(P)/_{Z_G}$ acts on $\mathcal{A}(P)$ and induces a free action on the subset $\mathcal{A}^*(P) \subset \mathcal{A}(P)$ of irreducible connections. Denote by $\mathcal{B}(P) := \mathcal{A}(P)/_{\mathrm{Aut}\,(P)}$ and $\mathcal{B}^*(P) := \mathcal{A}^*(P)/_{\mathrm{Aut}\,(P)}$ the corresponding quotients; these are the **moduli spaces** of gauge equivalence classes of (irreducible) connections on P [FU].

Now choose a **Riemannian metric** g on X, and let $* = *_g$ be the corresponding Hodge operator. The **Yang-Mills functional**

$$YM_X^g(P) : \mathcal{A}(P) \to \mathbb{R}_{\geq 0}$$

associates to a connection A the L^2-norm

$$\|F_A\|^2 := -\int_X \operatorname{tr}(F_A \wedge *F_A)$$

of its curvature. Since $-\operatorname{tr} : LG \otimes LG \to \mathbb{R}$ is an Ad-invariant metric, and since $F_{A \cdot f} = f^{-1} \circ F_A \cdot f$ for a gauge transformation f, $YM_X^g(P)$ descends to a well-defined functional on the moduli space $\mathcal{B}(P)$. A **Yang-Mills connection** is a critical point of $YM_X^g(P)$. The **Yang-Mills equations**, i.e. the Euler-Lagrange equations of the corresponding variational problem, read [FU]

$$D_A(*F_A) = 0.$$

These equations form a second order system of partial differential equations in the variable A and — together with the Bianchi identity — express the fact that the curvature of a Yang-Mills connection is harmonic with respect to its own Laplacian. The critical set of $YM_X^g(P)$ in $\mathcal{B}(P)$, i.e. the set of gauge equivalence classes of Yang-Mills connections, is the **Yang-Mills moduli space**. For an abelian structure group G Yang-Mills theory reduces to ordinary Hodge theory.

Example 1. Let $G = S^1$ and consider a bundle P over a (not necessarily simply-connected) 4-manifold X. A connection $A \in \mathcal{A}(P)$ is Yang-Mills if its curvature is harmonic, i.e. iff $-\frac{1}{2\pi i}F_A$ is the unique g-harmonic representative of the first Chern class $c_1(P) \in H^2_{DR}(X)$. Every other Yang-Mills connection is then of the form $A + i \cdot \alpha$ for a closed 1-form α. The gauge group in this abelian case is just the group $C^\infty(X; S^1)$ of differentiable maps from X to S^1 with the obvious action on $\mathcal{A}(P)$. The moduli space of Yang-Mills connections can therefore be identified with the quotient $H^1_{DR}(X)/_{H^1(X;\mathbb{Z})}$.

So far we have not used the assumption that our base manifold is of dimension 4. This 4-dimensional case has three important special features.

i) Conformal invariance: If one replaces the Riemannian metric g by a point-wise conformal metric $\lambda^2 \cdot g$, then the integrand $-\operatorname{tr}(F_A \wedge *F_A)$ of the Yang-Mills functorial does not change. This shows that the Yang-Mills functional $YM_X^g(P)$ depends only on the conformal class $[g]$ of the metric [FU].

ii) Self duality: The Hodge operator $*$ splits the bundle $\Lambda^2 = \Lambda^2 T_X^*$ of 2-forms on X into (± 1)-eigenbundles

$$\Lambda^2 = \Lambda_+ \oplus \Lambda_-,$$

the bundles of **self-dual** and **anti-self-dual** 2-forms. This splitting is orthogonal with respect to the Hodge inner product and induces a decomposition

$F_A = F_{A+} + F_{A-}$ of curvature forms [FU]. The self-dual and anti-self-dual components are characterized by the equations

$$*F_{A\pm} = \pm F_{A\pm}.$$

A connection $A \in \mathcal{A}(P)$ is called (anti-) self-dual if its curvature F_A is an (anti-) self-dual form.

iii) Topological constraints: By Chern-Weil theory the differential form $-\frac{1}{4\pi^2}\mathrm{tr}\,(F_A \wedge F_A)$ represents a characteristic class of the bundle P/X [FU], [K2]. The integral

$$-\int_X \mathrm{tr}\,(F_A \wedge F_A) = ||F_{A+}||^2 - ||F_{A-}||^2$$

depends therefore only on the bundle P and not on the connection A. Since $||F_A||^2 = ||F_{A+}||^2 + ||F_{A-}||^2$, one finds a **topological lower bound**

$$YM_X^g(P)(A) \geq \left|\int_X \mathrm{tr}\,(F_A \wedge F_A)\right|$$

for the Yang-Mills functional. To make this bound explicit, assume P is a $U(2)$-bundle with induced $PU(2)$-bundle $\bar{P} = P/_{S^1}$ corresponding to the exact sequence

$$1 \to S^1 \to U(2) \longrightarrow PU(2) \to 1.$$

The associated split exact sequence of Lie algebras

$$0 \to i\mathbb{R} \overset{\frac{1}{2}\mathrm{tr}}{\longrightarrow} LU(2) \to LPU(2) \to 0$$

leads to a decomposition

$$F_A = \frac{1}{2}\mathrm{tr}\,(F_A) \cdot \mathrm{id} + F_A^0$$

of the curvature form of a connection A into a **central and a trace free** component. The central part $\frac{1}{2}\mathrm{tr}\,(F_A) \cdot \mathrm{id}$ represents $-\pi i\, c_1(P)$ whereas the trace free component F_A^0 is the curvature of the induced $PU(2)$- connection $\bar{A}$ on $\bar{P}$ [D1]. Using

$$||F_A||^2 = \frac{1}{2}||\mathrm{tr}\,(F_A)||^2 + ||F_A^0||^2, ||F_A^0||^2 = ||F_{\bar{A}}||^2 = ||F_{\bar{A}+}||^2 + ||F_{\bar{A}-}||^2,$$

and

$$||F_{\bar{A}+}||^2 - ||F_{\bar{A}-}||^2 = \left[c_1(P)^2 - 4c_2(P)\right] \cdot 2\pi^2,$$

one obtains

$$YM_X^g(P)(A) = \frac{1}{2}||\mathrm{tr}\,(F_A)||^2 \pm \left[4c_2(P) - c_1(P)^2\right] \cdot 2\pi^2 + 2||F_{\bar{A}\pm}||^2.$$

The central and the trace free component of F_A may be minimized independently. The set of absolute minima of $YM_X^g(P)$ consists therefore of connections A with $\mathrm{tr}\,(F_A)$ harmonic and $F_{\bar{A}}$ self-dual or anti- self-dual, depending on the sign of $4\,c_2\,(P) - c_1\,(P)^2$.

2.2 Moduli spaces of instantons

A principal $PU(2)$-bundle $\bar{P}$ over X is — up to differentiable equivalence — classified by two characteristic classes, $w_2(\bar{P}) \in H^2(X;\mathbb{Z}/_2)$ and $p_1(\bar{P}) \in H^4(X;\mathbb{Z})$. To each integral lift $c \in H^2(X;\mathbb{Z})$ of $w_2(\bar{P})$ corresponds a principal $U(2)$-bundle P with $c_1(P) = c$ which induces $\bar{P}$, i.e. $\bar{P} \cong P/_{S^1}$, and thus $p_1(\bar{P}) = c_1(P)^2 - 4c_2(P)$. The bundle P can be chosen as $SU(2)$-bundle iff $w_2(\bar{P}) = 0$ [HH].
Suppose from now on that P is a principal $SU(2)$-bundle with $c_2(P) \geq 0$ or a principal $PU(2)$-bundle with $p_1(P) \leq 0$. An **instanton** on P — relative to the Riemannian metric g — is a connection $A \in \mathcal{A}(P)$ which is anti-self-dual with respect to g, i.e. $*_g F_A = -F_A$.
Instantons are clearly Yang-Mills connections; they give absolute minima of the Yang-Mills functional if they exist.
Furthermore, a $U(2)$-connection A minimizes the Yang-Mills functional iff $\mathrm{tr}(F_A)$ is harmonic and the induced $PU(2)$-connection $\bar{A}$ is an instanton. The gauge group $\mathrm{Aut}(P)$ leaves the subset $\mathcal{A}_-(P) := \{A \in \mathcal{A}(P) \mid *_g F_A = -F_A\}$ of instantons invariant.
The **moduli space** of **g-instantons** on P is defined as the orbit space [FU]

$$\mathcal{M}^g_X(P) := \mathcal{A}_-(P)/_{\mathrm{Aut}(P)}.$$

Denote by $\mathcal{M}^g_X(P)^* \subset \mathcal{B}^*(P)$ the subspace of irreducible instantons. This space admits another description: Let $F_+ : \mathcal{A}^*(P) \to A^2_+(ad(P))$ be the function $A \mapsto F_{A+}$; since F_+ is an $\mathrm{Aut}^*(P)$-equivariant map, it induces a section of the vector bundle

$$\begin{array}{c} \mathcal{A}^*(P) \times_{\mathrm{Aut}^*(P)} A^2_+(ad(P)) \\ \downarrow \uparrow \; F_+ \\ \mathcal{B}^*(P) \end{array}$$

over $\mathcal{B}^*(P)$. Its zero-set $\{[A] \in \mathcal{B}^*(P) \mid F_{A+} = 0\}$ is the moduli space $\mathcal{M}^g_X(P)^*$ of irreducible instantons on P.

We will now discuss local and global properties of these moduli spaces.

i) Local structure: Associated with any instanton A on P there is the following **fundamental elliptic complex:**

$$(*) \qquad 0 \to A^0(ad(P)) \xrightarrow{D_A} A^1(ad(P)) \xrightarrow{D_{A+}} A^2_+(ad(P)) \to 0.$$

The cohomology groups H^i_A $i = 0,1,2$ of this complex are finite dimensional vector spaces and have the following significance [L]: $A^0(ad(P))$ is the tangent space of the gauge group at the identity 1_P. Viewing $A^1(ad(P))$ as tangent space of $\mathcal{A}(P)$ at A one can identify the operator $D_A : A^0(ad(P)) \to A^1(ad(P))$ with the differential of the orbit map $A\cdot : \mathrm{Aut}(P) \to \mathcal{A}(P)$ at 1_P.
Its kernel H^0_A is therefore the tangent space of the isotropy group $\mathrm{Aut}(P)_A$

at the identity, whereas $\mathrm{Im}(D_A)$ corresponds to the tangent space of the orbit through A.
Another connection $A+\alpha, \alpha \in \mathcal{A}^1(ad(P))$ is also an instanton iff it satisfies the anti-self-duality equation $(F_{A+\alpha})_+ = 0$, i.e.

$$D_{A+}(\alpha) + [\alpha \wedge \alpha]_+ = 0.$$

The linearization $D_{A+}(\alpha) = 0$ of this equation describes the space of tangents to curves in $\mathcal{A}_-(P)$ through A. Thus the first cohomology H^1_A of the fundamental complex $(*)$ becomes the 'Zariski' tangent space to $\mathcal{M}^g_X(P)$ at $[A]$. If $\mathrm{Aut}^*(P)_A = \{1\}$, and if $H^2_A = 0$, then $\mathcal{M}^g_X(P)$ is a manifold near the point $[A]$ with tangent space H^1_A. In general one has the following **Kuranishi description** for the germ of the space $\mathcal{M}^g_X(P)$ at $[A]$.

Theorem 1 [AHS], [L] . *Let A be an instanton on P. There exists an $\mathrm{Aut}^*(P)_A$-invariant neighborhood U of $0 \in H^1_A$ and an $\mathrm{Aut}^*(P)_A$-equivariant differentiable map*

$$K_{[A]} : U \to H^2_A$$

with $K_{[A]}(0) = 0$, such that $(K^{-1}_{[A]}(0)/_{\mathrm{Aut}^*(P)_A}, 0)$ describes the germ of $\mathcal{M}^g_X(P)$ at $[A]$.

The proof of this result is an adaption of arguments which Kuranishi used to construct deformations of complex structures on complex manifolds [Ku].

Example 2. Let P be a non-flat $PU(2)$-bundle over X. A connection A is **reducible** iff $\mathrm{Aut}(P)_A$ is non-trivial, in which case one has $\mathrm{Aut}(P)_A \cong S^1$ [FU]. This means there exists a unitary line bundle L and a S^1-connection D_L on L such that $ad(P)$ and D_A split in the form $ad(P) \cong L \oplus \epsilon, D_A = D_L \oplus d$, where d is the product connection on the trivial real line bundle ϵ. If A is an instanton, then also the fundamental complex splits:

$$\begin{array}{ccccccccc} 0 & \to & A^0(L) & \xrightarrow{D_L} & A^1(L) & \xrightarrow{D_{L+}} & A^2_+(L) & \to 0 \\ & & \oplus & & \oplus & & \oplus & \\ 0 & \to & A^0 & \xrightarrow{d} & A^1 & \xrightarrow{d_+} & A^2_+ & \to 0. \end{array}$$

Suppose for simplicity D_{L+} is onto, so that $H^2_A = \mathbb{H}^2_+(X)$ is the space of self-dual harmonic 2-forms on X.
The action of $\mathrm{Aut}(P)_A \cong S^1$ on the fundamental complex is trivial on the lower part and induces a complex structure on H^1_L. The Kuranishi map

$$K_{[A]} : H^1_L \to \mathbb{H}^2_+(X)$$

is S^1-invariant and thus factors through the cone over the complex projective space $\mathbb{P}(H^1_L)$:

$$\hat{K}_{[A]} : c\mathbb{P}(H^1_L) \to \mathbb{H}^2_+(X).$$

The zero-set of $\hat{K}_{[A]}$ near the cone point is a local model for $\mathcal{M}^g_X(P)$ near the reducible instanton $[A]$.

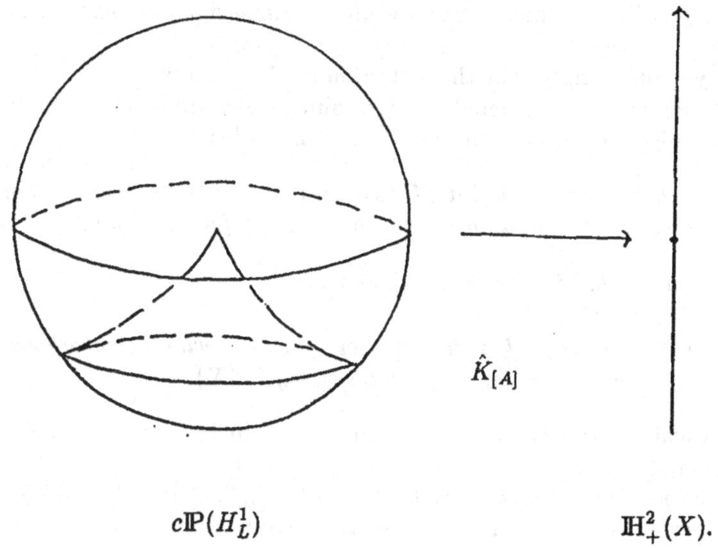

The Kuranishi model $(K^{-1}_{[A]}(0)/_{\mathrm{Aut}^*(P)_A}, 0)$ for the instanton moduli space shows that the **expected dimension** for $\mathcal{M}^g_X(P)$ at the point $[A]$ is equal to $\dim H^1_A - \dim H^2_A - \dim H^0_A$, i.e. it is minus the index of the fundamental elliptic complex $(*)$. Using the Atiyah-Singer index theorem to calculate this index one finds

$$-2p_1(ad(P)) - 3(1 + b_+(X)),$$

where $b_+(X)$ is the dimension of the space $\mathbb{H}^2_+(X)$ [FU].

ii) Global properties: The group $\mathrm{Aut}^*(P) = \mathrm{Aut}\,(P)/_{Z_G}$ acts freely on the set $\mathcal{A}^*(P)$ of irreducible connections. Using the obvious action on $P/_{Z_G}$ one can form a bundle

$$\begin{array}{c} \mathcal{A}^*(P) \times_{\mathrm{Aut}^*(P)} P/_{Z_G} \\ \downarrow \\ \mathcal{B}^*(P) \times X \end{array}$$

over the product $\mathcal{B}^*(P) \times X$. The restriction of this bundle $\mathbb{P}$ to $\mathcal{M}^g_X(P)^* \times X$ is the **universal bundle** $\bar{\mathbb{P}}$; over the slice $[A] \times X$ it is isomorphic to $P/_{Z_G}$.

So far we have treated the moduli spaces $\mathcal{M}^g_X(P) \subset \mathcal{B}(P)$ as topological spaces without explaining their topology.
In order to make this point precise, one has to use suitable **Sobolev norms** for $\mathcal{A}(P)$ and $\mathrm{Aut}(P)$, so that the completed spaces become Banach spaces and Banach Lie groups respectively [FU], [L]. Then one proves a slice theorem for the action of this infinite dimensional Lie group, and shows the orbit

map $\mathcal{A}^*(P) \to \mathcal{B}^*(P)$ is the projection of an infinite dimensional, differentiable principal bundle.
Elliptic regularity results imply that the instanton moduli spaces $\mathcal{M}^g_X(P)^* \subset \mathcal{M}^g_X(P)$ can be considered as finite dimensional differentiable subspaces in these infinite dimensional spaces $\mathcal{B}^*(P) \subset \mathcal{B}(P)$ [FU].

Theorem 2 [FU]. *Let P be a non-flat $SU(2)$- or $PU(2)$-bundle over X. For an open and dense set of Riemannian metrics g on X the following holds:*

i) $\mathcal{M}^g_X(P)^*$ is a manifold of the expected dimension.
ii) $\mathcal{M}^g_X(P) = \mathcal{M}^g_X(P)^*$ if $b_+(X) > 0$.
iii) $\mathcal{M}^g_X(P) \setminus \mathcal{M}^g_X(P)^*$ consists of a discrete set of points with neighborhoods equivalent to cones over complex projective space if $b_+(X) = 0$.

This result is basically a transversality theorem, whose main ingredient is the Sard-Smale theorem [Sm].
The reason for ii) is that the subspace $\mathbb{H}^2_-(X) \subset H^2_{DR}(X)$ of anti-self-dual 2-forms is a proper subspace which for a generic metric avoids a given set of integral classes [FU]. The neighborhood structure in case iii) follows from the Kuranishi description [FU].

The next two results of K. Uhlenbeck are crucial for studying compactness properties of instanton moduli spaces.

Theorem 3 [U2]. *Let $\{A_\nu\}$ be a sequence of instantons on P. Then either*

i) there is a subsequence $\{A_{\nu'}\}$ and a sequence $\{\tilde{A}_{\nu'}\}$ of gauge equivalent connections converging to an instanton on P, or
ii) there exist finitely many points $x_1, \ldots, x_m \in X$, a subsequence $\{A_{\nu'}\}$ and a sequence $\{\tilde{A}_{\nu'}\}$ of gauge equivalent connections which converges to an anti-self-dual connection on the restricted bundle $P|_{X \setminus \{x_1,\ldots,x_m\}}$ over $X \setminus \{x_1, \ldots, x_m\}$.

Theorem 4 [U1]. *Let A_0 be an anti-self-dual connection on a $SU(2)$- or $PU(2)$-bundle P_0 over $X \setminus \{x_1, \ldots, x_m\}$ with finite Yang-Mills action $\|F_{A_0}\|^2 < \infty$. Then there exists a gauge transformation $f \in \mathrm{Aut}(P_0)$, so that f^*P_0 and f^*A_0 extend smoothly to a bundle and an instanton over X.*

The latter theorem is actually a local result which applies more generally to Yang-Mills connections with finite energy, i.e. connections A_0 with $D_{A_0}(*F_{A_0}) = 0$ and $\|F_{A_0}\|^2 < \infty$ [U1]. It leads to a natural **compactification** of $\mathcal{M}^g_X(P)$ by **ideal instantons** [D2]. Heuristically this means: If a sequence $\{[A_\nu]\}$ in $\mathcal{M}^g_X(P)$ has no convergent subsequence then a subsequence $\{[\tilde{A}_{\nu'}]\}$ can be found which 'converges' to a point $([A'], \{x_1, \ldots, x_m\}) \in \mathcal{M}^g_X(P') \times S^m X$. Here P' is a bundle with $w_2(P') = w_2(P)$ and $p_1(P') \geq p_1(P)$, $S^m X$ is the m-th symmetric product of X with $m = \frac{1}{4}(p_1(P') - p_1(P))$. These limiting ideal connections

$([A'], \{x_1, \ldots, x_m\})$ are the boundary points of a compactification of $\mathcal{M}_X^g(P)$. More precisely, there is a topology on the union

$$\bigcup_{0 \leq m \leq [\frac{-p_1}{4}]} \mathcal{M}_X^g(P') \times S^m X,$$

so that the closure $\bar{\mathcal{M}}_X^g(P)$ of $\mathcal{M}_X^g(P)$ is compact [D2]. In particular, $\mathcal{M}_X^g(P)$ is already compact if $p_1(P) \in \{0, -1, -2, -3\}$ or if $p_1(P) = -4$ and $w_2(P) \neq 0$ [FS1]. The orientability of the instanton spaces has been shown by Donaldson.

Theorem 5 [D5]. *The moduli spaces $\mathcal{M}_X^g(P)^*$ are orientable for a generic metric g. An orientation is given by the choice of an orientation of $\mathbb{H}_+^2(X)$ and the choice of an integral lift c of $w_2(P)$ if $w_2(P) \neq 0$.*

Let $c \in H^2(X; \mathbb{Z})$ be an integral lift of $w_2(P)$. Fixing c and choosing the opposite orientation for $\mathbb{H}_+^2(X)$ reverses the orientation on $\mathcal{M}_X^g(P)^*$. Fixing the orientation of $\mathbb{H}_+^2(X)$ and choosing another lift $c' = c + 2\ell$ of $w_2(P)$ reverses the orientation of $\mathcal{M}_X^g(P)^*$ iff the self-intersection number ℓ^2 of the class $\ell \in H^2(X; \mathbb{Z})$ is odd [D4].

So far we do not know on which bundles P/X instantons actually **exist**. For principal $SU(2)$-bundles C. Taubes has proved a general existence theorem.

Theorem 6 [T]. *Let P be a $SU(2)$-bundle with second Chern class $c_2 \geq 0$ over closed oriented Riemannian 4-manifold X. Then*

$$\mathcal{M}_X^g(P)^* \neq \emptyset$$

for $c_2 \geq \max(\frac{4}{3} b_+(X), 1)$ if $b_+(X) \neq 2$, and for $c_2 \geq 4$ if $b_+(X) = 2$.

There are weaker numerical conditions if g is a **generic** metric on a manifold with $b_+(X) \notin \{0, 1, 3\}$. In this case Taubes shows that irreducible instantons exist on all $SU(2)$-bundles with $c_2 \geq b_+(X)$ [T].

3 Donaldson invariants

This section contains a brief description of Donaldson's method for constructing differential invariants from moduli spaces of instantons.

Let X be a closed oriented simply-connected differentiable 4-manifold with $b_+(X) \equiv 1 \pmod 2$. Fix a non-flat principal bundle P over X with structure group $SU(2)$ or $PU(2)$ and $p_1(P) \leq 0$. The first Pontrjagin class $p_1(\mathbb{P}) \in H^4(\mathcal{B}^*(P) \times X; \mathbb{Z})$ of the universal bundle $\mathbb{P}/\mathcal{B}^*(P) \times X$ induces a homomorphism

$$\mu : H_2(X; \mathbb{Z}) \to H^2(\mathcal{B}^*(P); \mathbb{Z})$$

which sends a class $C \in H_2(X; \mathbb{Z})$ to $\mu(C) := p_1(\mathbb{P})/C$.
Choose a generic metric g in X and an orientation o of the instanton moduli space

$\mathcal{M}_X^g(P)^* \subset \mathcal{B}^*(P)$. Then $\mathcal{M}_X^g(P)^*$ becomes an oriented manifold of dimension $2d$ with

$$d = -p_1(P) - \frac{3}{2}(1 + b_+(X)) \text{ [D4]}.$$

If these choices would define a **fundamental class** $[\mathcal{M}_X^g(P)^*] \in H_{2d}(\mathcal{B}^*(P); \mathbb{Z})$, then one could construct a homogeneous polynomial function

$$\gamma_{X,o}^g(P) : S^d H_2(X; \mathbb{Z}) \to \mathbb{Z}$$

on $H_2(X; \mathbb{Z})$ by assigning the value

$$< \mu(C_1) \cup \ldots \cup \mu(C_d), [\mathcal{M}_X^g(P)^*] >$$

to $(C_1, \ldots, C_d)$.

In order to use this idea for the construction of differential invariants of X one has to deal with the following problems [D4]:

I) The instanton spaces $\mathcal{M}_X^g(P)^*$ are generally **non-compact** and therefore do not define fundamental classes in the usual sense.
II) The definition of $\mathcal{M}_X^g(P)^*$ depends on the **choice** of a **Riemannian structure** on X, not only on the differential topology.

The first problem can be solved for all $PU(2)$-bundles with $w_2(P) \neq 0$ and also for $SU(2)$-bundles in the **stable range**

$$4\, c_2(P) \geq 3(1 + b_+(X)) + 2.$$

Outside of this range the definition of $\gamma_{X,o}^g(P)$ has to be modified.
The second problem is more serious; it leads to a basic distinction between the special case $b_+(X) = 1$ and the general situation $b_+(X) > 1$. If $b_+(X) > 1$, then the polynomials $\gamma_{X,o}^g(P)$ are actually independent of the metric and therefore yield invariants of the differential structure of X [D4]. In the special case $b_+(X) = 1$, however, they do depend on g. This dependence has to be taken into account and leads to more complicated C^∞-invariants [D3], [D6].

3.1 Definition of the polynomials

Let $\bar{\mathcal{M}}_X^g(P)$ be the **Donaldson compactification** of the moduli space $\mathcal{M}_X^g(P)^*$ by ideal instantons. This compactification is a **stratified** space with $\mathcal{M}_X^g(P)^*$ as open stratum; it carries a fundamental class $[\bar{\mathcal{M}}_X^g(P)]$ if the lower strata have codimension at least 2, i.e. if

$$\dim(\bar{\mathcal{M}}_X^g(P) \setminus \mathcal{M}_X^g(P)^*) \leq \dim(\mathcal{M}_X^g(P)^*) - 2.$$

The dimension of a stratum

$$\bar{\mathcal{M}}_X^g(P) \cap (\mathcal{M}_X^g(P') \times S^m X)$$

corresponding to a **non-flat** bundle P' with $p_1(P') > p_1(P)$ and $m = \frac{1}{4}(p_1(P') - p_1(P))$ is at most equal to

$$-2p_1(P') - 3(1 + b_+(X)) + 4m.$$

The codimension of such a stratum is therefore at least 4. If P' is **flat** however, then the corresponding stratum is of dimension $-p_1(P)$. Since $w_2(P') = w_2(P)$ and $\pi_1(X) = 0$, a stratum associated to a flat bundle P' can only occur if $w_2(P) = 0$, i.e. in the $SU(2)$-case. Thus in this case one needs the condition $-p_1(P) \leq -2p_1(P) - 3(1+b_+(X)) - 2$ to guarantee the existence of a fundamental class $[\bar{\mathcal{M}}^g_X(P)]$. This condition defines the stable range mentioned above [D7].
Consider now a homology class $C \in H_2(X;\mathbb{Z})$, and let L_C be the complex line bundle over $\mathcal{M}^g_X(P)^*$ which is defined by the restriction of the class $\mu(C) \in H^2(\mathcal{B}^*(P);\mathbb{Z})$ to $\mathcal{M}^g_X(P)^*$. These line bundles have natural extensions to line bundles $\bar{L}_C$ over the compactification $\bar{\mathcal{M}}^g_X(P)$ [D7]. Sending $C_1, \ldots, C_d \in H_2(X;\mathbb{Z})$ to

$$< c_1(\bar{L}_{C_1}) \cup \ldots \cup c_1(\bar{L}_{C_d}), [\bar{\mathcal{M}}^g_X(P)] >$$

yields therefore a well-defined polynomial function

$$\gamma^g_{X,o}(P) : S^d H_2(X;\mathbb{Z}) \to \mathbb{Z}$$

for $PU(2)$-bundles with $w_2 \neq 0$ and for $SU(2)$-bundles in the stable range $4c_2 \geq 3(1 + b_+(X)) + 2$.

For certain $SU(2)$-bundles with $4c_2 < 3(1 + b_+(X)) + 2$ there is another way to make sense of the expression $< \mu(C_1) \cup \ldots \cup \mu(C_d), -\mathcal{M}^g_X(P)^* - >$. The idea is to compactify $\mathcal{M}^g_X(P)^*$ by **truncating** its ends and then introducing a **correction term** at infinity.
This method has been used by Donaldson in the construction of his Γ-invariant [D3]. His construction works for $SU(2)$-bundles P with $c_2(P) = 1$ over 4-manifolds with $b_+(X) = 1$ and $b_-(X) > 0$.
The instanton moduli space $\mathcal{M}^g_X(P)^*$ in this case will be smooth of dimension 2, if the metric is generic, but it will usually not be compact. Its end can be described in the following way [D3]: Let $\omega_g \in \mathbb{H}^2_+(X)$ be a non-zero self-dual 2-form and denote the zero-set of the corresponding section in Λ_+ by $Z(\omega_g)$. This zero-set is a smooth curve which depends on g but not on the choice of ω_g. Donaldson shows that there is a function

$$\tau : X \times (0,1) \to \mathcal{B}^*(P)$$

mapping $Z(\omega_g) \times (0,1)$ diffeomorphically onto an open subset of $\mathcal{M}^g_X(P)^*$, so that the complement $\mathcal{M}^g_X(P)^* \backslash \tau(Z(\omega_g) \times (0,1))$ is compact.

In other words, the end of $\mathcal{M}^g_X(P)^*$ looks like $Z(\omega_g) \times (0,1)$. The truncated moduli space

$$\hat{\mathcal{M}}^g_X(P) := \mathcal{M}^g_X(P)^* \backslash \tau(Z(\omega_g) \times (0,1))$$

is a compact surface with boundary $\tau_1(Z(\omega_g))$, so that the choice of an orientation defines a relative fundamental class $[\hat{M}^g_X(P)] \in H_2(\mathcal{B}^*(P), \tau_1(Z(\omega_g)); \mathbb{Z})$. Let $e \in H_2(X, Z(\omega_g); \mathbb{Z})$ be the Lefschetz dual of the Euler class of the bundle $\Lambda_+/_{\Lambda^0 \cdot \omega_g}$ over $X \setminus Z(\omega_g)$. The difference $2[\hat{\mathcal{M}}^g_X(P)] - \tau_{1*}(e)$ maps to zero in $H_1(\tau_1(Z(\omega_g)); \mathbb{Z})$ and hence comes from a uniquely determined natural class

$$[\mathcal{M}^g] \in H_2(\mathcal{B}^*(P); \mathbb{Z}).$$

The evaluation of $\mu(C)$ on $[\mathcal{M}^g]$ defines a linear polynomial

$$\gamma^g_{X,o}(P) : H_2(X; \mathbb{Z}) \to \mathbb{Z}.$$

3.2 Dependence on the metric

Suppose now that we have well-defined polynomials

$$\gamma^g_{X,o}(P) : S^d H_2(X; \mathbb{Z}) \to \mathbb{Z}$$

for generic metrics g.
In order to construct differential invariants from these metric dependent polynomials one has to analyze the behaviour of instanton moduli space $\mathcal{M}^{g_t}_X(P)$ in 1-parameter families of metrics g_t [D4].
If $b_+(X) > 1$, then any path $(g_t)_{t\in[0,1]}$ of metrics with generic end points g_0, g_1 can be approximated arbitrarily closely by a path $(\tilde{g}_t)_{t\in[0,1]}$ with the same endpoints, such that the following holds [D4]: For any non-flat bundle P', which contributes to the Donaldson compactification $\bar{\mathcal{M}}^g_X(P)$, the family $\mathcal{M}^{\tilde{g}_t}_X(P')$ is contained in $\mathcal{B}^*(P')$ and defines a smooth manifold

$$\mathcal{M}^{\tilde{g}}_X(P') := \bigcup_{t\in[0,1]} \mathcal{M}^{\tilde{g}_t}_X(P') \times \{t\} \subset \mathcal{B}^*(P') \times [0,1]$$

in $\mathcal{B}^*(P') \times [0,1]$ with boundary

$$\partial \mathcal{M}^{\tilde{g}}_X(P') = \mathcal{M}^{g_0}_X(P') \amalg -\mathcal{M}^{g_1}_X(P').$$

Using these bordisms Donaldson concludes $\gamma^{g_0}_{X,o}(P) = \gamma^{g_1}_{X,o}(P)$, i.e. the polynomials $\gamma^g_{X,o}(P)$ — defined for $PU(2)$-bundles, or for $SU(2)$-bundles in the stable range — are independent of the metric if $b_+(X) > 1$ [D4].

Recall that an orientation o of $\mathcal{M}^g_X(P)^*$ is given by the choice of an orientation of $\mathbb{H}^2_+(X)$ and the choice of an integral lift c of $w_2(P)$. Donaldson defines a H_+**-orientation** as an orientation of a maximal positive subspace $H_+ \subset H^2_{DR}(X)$ for the intersection form. Since the space of all such positive subspaces is contractible, the choice of a particular H_+ is not essential.

Theorem 7 [D4]. *Let X be a closed simply-connected oriented differentiable 4-manifold with $b_+(X) = 2p+1 > 1$.*
Let P/X be a non-flat principal $PU(2)$-bundle with $w_2(P) \neq 0$ or a non-flat principal $SU(2)$-bundle with $4c_2(P) \geq 3(1 + b_+(X)) + 2$. Associated to any H_+-orientation o_+ and any integral lift c of $w_2(P)$ if $w_2(P) \neq 0$ there exists a polynomial

$$\gamma_{X,c,o_+}(P) : S^d H_2(X; \mathbb{Z}) \to \mathbb{Z}$$

of degree $d = -p_1(P) - 3(1+p)$ with the following properties:

i) $\gamma_{X,c,-o_+}(P) = -\gamma_{X,c,o_+}(P)$.
ii) $\gamma_{X,c+2\ell,o_+}(P) = \epsilon(\ell)\gamma_{X,c,o_+}(P)$ *where*

$$\epsilon(\ell) = \begin{cases} 1 & \text{for} \quad \ell^2 \equiv 0 \pmod 2 \\ -1 & \text{for} \quad \ell^2 \equiv 1 \pmod 2). \end{cases}$$

iii) If $f : X' \to X$ is an orientation preserving diffeomorphism, then $\gamma_{X',f^*c,f^*o_+}(f^*P) = f^*\gamma_{X,c,o_+}(P)$.

The definition of differential invariants for 4-manifolds X with $b_+(X) = 1$ is more difficult. In this case new phenomena occur due to the appearance of **reducible** connections in generic 1-parameter families $\mathcal{M}_X^{\tilde{g}_t}(P')$ of instanton moduli spaces. A transversality argument shows that there will usually exist finitely many values of t for which some of the moduli spaces $\mathcal{M}_X^{\tilde{g}_t}(P')$ are not entirely contained in $\mathcal{B}^*(P')$. If this happens at a time t_0 for a bundle P' which is needed for the Donaldson compactification of $\mathcal{M}_X^{g}(P)^*$, then the polynomials $\gamma_{X,o}^{\tilde{g}_t}(P)$ will, in general, be different for $t < t_0$ and $t > t_0$ [D6], [Mo1].
A moduli space $\mathcal{M}_X^{\tilde{g}_t}(P')$ contains the class of a reducible connection iff there exists an integral lift c' of $w_2(P')$ with $c'^2 = p_1(P')$ which is contained in the subspace $\mathbb{H}^2_-(X) \subset H^2_{DR}(X)$ of $\tilde{g}_t$-anti-self-dual 2-forms. Therefore one proceeds as follows [Mo1]: The **positive cone**

$$\Omega_X := \{\alpha \in H^2_{DR}(X) \mid \alpha \cdot \alpha > 0\}$$

of a manifold X with $b_+(X) = 1$ consists of two components, $\Omega_X = \Omega_+ \cup \Omega_-$, with $\Omega_- = -\Omega_+$. The choice of a H_+-orientation is equivalent to the choice of one of these components.
For every integral class $E \in H^2(X; \mathbb{Z})$ with $w_2(P) = \bar{E} \pmod 2$ and $p_1(P) \leq E^2 < 0$ let

$$W_E := \{\alpha \in \Omega_X \mid \alpha \cdot E = 0\}$$

be the corresponding **wall** in Ω_X. Denote the set of connected components of

$$\Omega_X \setminus \bigcup_E W_E$$

by $\mathcal{C}_X(P)$; this is the set of **chambers**.
The metric dependent polynomials $\gamma_{X,c,o_+}^{g}(P)$ do not change as long as the

intersection of $\mathbb{H}^2_+(X)$ (defined with respect to g) with the chosen component of Ω_X stays in a fixed chamber.

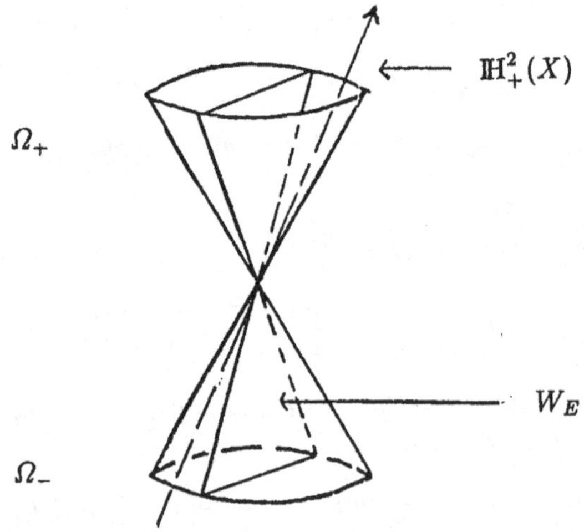

The main difficulty is now to find a **universal formula**, i.e. a formula which does not depend on the C^∞-structure, for the difference $\gamma^{g'}_{X,c,o_+}(P) - \gamma^{g}_{X,c,o_+}(P)$ of polynomials corresponding to adjacent chambers [Mo1].
If that is possible, then a differential invariant can be constructed by extending $\gamma^{g}_{X,c,o_+}(P)$ to a function

$$\Gamma_{X,c}(P) : \mathcal{C}_X(P) \to \mathrm{Hom}_{\mathbb{Z}}(S^d H_2(X;\mathbb{Z}), \mathbb{Z}).$$

We will need two special cases in which this idea has been worked out.
The first case is Donaldson's **Γ-invariant** which historically was the first differential invariant to be constructed [D3]. It corresponds to a $SU(2)$-bundle P with $c_2(P) = 1$ over an indefinite 4-manifold X with $b_+(X) = 1$. The set $\mathcal{C}_X$ of chambers in this situation consists of the components of $\Omega_X \setminus \bigcup_{E^2=-1} W_E$.

Theorem 8 [D3]. *There exists a function*

$$\Gamma_X : \mathcal{C}_X \to H^2(X;\mathbb{Z})$$

with the following properties:

i) $\Gamma_X(-C) = -\Gamma_X(C)$.
ii) If C_+, C_- are chambers in Ω_+, then $\Gamma_X(C_+) - \Gamma_X(C_-) = 2\sum E$ where the summation is over all classes $E \in H^2(X;\mathbb{Z})$ with $E^2 = -1$ and $C_- \cdot E < 0 < C_+ \cdot E$.
iii) $\Gamma_{X'} \circ f^* = f^* \circ \Gamma_X$ *for an orientation preserving diffeomorphism* $f : X' \to X$.

Note that the Γ-invariant is an invariant outside of the stable range $c_2 \geq 2$; its definition uses the truncation method alluded to earlier. The $PU(2)$-analog of the Γ-invariant is much easier to construct [OV2]; it corresponds to an irreducible $PU(2)$-bundle P with $w_2(P) \neq 0$ and $p_1(P) = -4$. In this case the chamber structure of Ω_X becomes trivial, i.e. there are only the two chambers Ω_+ and Ω_-.

Theorem 9 [OV2]. *Let P be an irreducible $PU(2)$-bundle with $w_2(P) \neq 0$ and $p_1(P) = -4$. Every integral lift c of $w_2(P)$ defines a map*

$$\bar{\Gamma}_{X,c} : \{\Omega_+, \Omega_-\} \to H^2(X; \mathbb{Z})$$

with the following properties:

i) $\bar{\Gamma}_{X,c}(\Omega_-) = -\bar{\Gamma}_{X,c}(\Omega_+)$.
ii) $\bar{\Gamma}_{X,c+2\ell} = \epsilon(\ell)\bar{\Gamma}_{X,c}$ *where*

$$\epsilon(\ell) = \begin{cases} 1 & \text{for} \quad \ell^2 \equiv 0 \pmod 2 \\ -1 & \text{for} \quad \ell^2 \equiv 1 \pmod 2. \end{cases}$$

iii) $\bar{\Gamma}_{X',f^*c} \circ f^* = f^* \circ \bar{\Gamma}_{X,c}$ *for an orientation preserving diffeomorphism* $f : X' \to X$.

For other situations in which invariants for manifolds with $b_+(X) = 1$ have been defined we refer to [Kot] and [Mo2].

At this point it is by no means clear that at least one of all the invariants which we have just described is actually **non-trivial**. In fact, the only result about these invariants which can be proved more or less from first principles is a **vanishing theorem** for connected sums.

Theorem 10 [D4]. *All polynomial invariants vanish for manifolds X which are smooth connected sums $X = X_1 \# X_2$ with each $b_+(X_i) > 0$.*

4 Hermitian-Einstein structures and stable bundles

In this section we treat the integrable case of instantons over algebraic surfaces and explain the relation of instanton moduli spaces to moduli spaces of stable algebraic vector bundles.

4.1 Hermitian-Einstein bundles

Let X be a smooth connected compact complex surface and let E be a differentiable complex vector bundle of rank r over X with $c_1(E) \in NS(X)$.
Denote by $A^{p,q}$ the bundle of (p,q)-forms and by $A^{p,q}(E)$ the space of sections in $A^{p,q} \otimes E$. A holomorphic structure $\mathcal{E}$ on E induces a differential operator

$$\bar{\partial}_{\mathcal{E}} : A^0(E) \to A^{0,1}(E)$$

satisfying the $\bar\partial$-Leibniz rule $\bar\partial_{\mathcal{E}}(\varphi \cdot s) = \bar\partial\varphi \otimes s + \varphi \cdot \bar\partial_{\mathcal{E}}(s)$, such that (using the same symbol for its extension to $A^{p,q}(E)$) $\bar\partial^2_{\mathcal{E}} = 0$ [K2]. Conversely, every **semi-connection** $\bar\partial_A$ on E, i.e. every differential operator

$$\bar\partial_A : A^0(E) \to A^{0,1}(E)$$

satisfying the $\bar\partial$-Leibniz rule, which is **integrable** in the sense that $\bar\partial^2_A = 0$, induces a holomorphic structure $\mathcal{E}_A$ on E such that $\bar\partial_{\mathcal{E}_A} = \bar\partial_A$. This is the vector bundle version of the Newlander-Nirenberg theorem [K2]. Let $\bar{\mathcal{A}}(E) = \bar\partial_{A_0} + A^{0,1}(\mathrm{End}\, E)$ be the affine space of semi-connections and denote by $\mathcal{H}(E) := \{\bar\partial_A \in \bar{\mathcal{A}}(E) \mid \bar\partial^2_A = 0\}$ the subspace of **holomorphic structures.** Two holomorphic structures $\bar\partial_A, \bar\partial_{A'}$ define holomorphically equivalent bundles $\mathcal{E}_A, \mathcal{E}_{A'}$ iff there exists a complex gauge transformation $f \in GL(E)$ with $\bar\partial_{A'} \circ f = f \circ \bar\partial_A$ [K2]. The natural action of $GL^*(E) := GL(E)/_{\mathbb{C}^* \cdot \mathrm{id}_E}$ on $\mathcal{H}(E)$ restricts to a free action on the subset $\mathcal{S}(E) := \{\bar\partial_A \in \mathcal{H}(E) \mid H^0(X; \mathrm{End}\, \mathcal{E}_A) = \mathbb{C} \cdot \mathrm{id}\}$ of **simple holomorphic structures.** The quotient

$$\mathcal{M}^s_X(E) := \mathcal{S}(E)/_{GL^*(E)}$$

is the **moduli space** of **simple holomorphic structures.**

Theorem 11 [I 2], [LO], [K2]. *$\mathcal{M}^s_X(E)$ is a locally Hausdorff complex space of finite dimension.*

The natural complex structure on $\mathcal{M}^s_X(E)$ is not reduced in general [LO]. Local models can be constructed as zero sets of germs of analytic maps

$$K_{[\bar\partial_A]} : H^1(X; \mathrm{End}\, \mathcal{E}_A) \to H^2(X; \mathrm{End}\, \mathcal{E}_A).$$

This moduli space will be used later to compare the moduli spaces of instantons and stable bundles.

Suppose now X admits a **Kähler metric** g with Kähler form ω.
Choose a Hermitian metric h on E and let P be the corresponding $U(r)$-bundle. The map $\mathcal{A}(P) \to \bar{\mathcal{A}}(E)$, sending a unitary connection A to the $(0,1)$-component of the operator $d_A : A^0(E) \to A^1(E)$, is a bijection and identifies the subset $\mathcal{H}(E) \subset \bar{\mathcal{A}}(E)$ of holomorphic structures on E with the set

$$\mathcal{A}^{1,1}(P) := \{A \in \mathcal{A}(P) \mid F_A \in A^{1,1}(\mathrm{End}\, E)\}$$

of **Chern connections** .
The Kähler metric g on the other hand splits the bundle $\Lambda^2_{\mathbb{C}}$ of $\mathbb{C}$-valued 2-forms into an orthogonal sum

$$\Lambda^2_{\mathbb{C}} = \Lambda_+ \otimes \mathbb{C} \oplus \Lambda_- \otimes \mathbb{C}.$$

One checks that

$$\Lambda_+ \otimes \mathbb{C} = \Lambda^{2,0} \oplus \Lambda^{0,2} \oplus \Lambda^0_{\mathbb{C}} \cdot \omega,$$

so that $\Lambda_- \otimes \mathbb{C} = \Lambda_\perp^{1,1}$ consists of **primitive** forms, i.e. (1,1)-forms orthogonal to ω [I 1]. Let

$$F_A = F_A^{2,0} + F_A^{0,2} + \hat{F}_A \cdot \omega + F_{A\perp}^{1,1}$$

be the corresponding decomposition of curvature forms.
The **Ricci curvature** of a connection A is the (real) endomorphism $i \cdot \hat{F}_A$. A connection $A \in \mathcal{A}^{1,1}(P)$ is a **Hermitian-Einstein** (H.-E.) connection if its Ricci curvature is constant,

$$i\hat{F}_A = c \cdot \mathrm{id}_E.$$

The bundle E is an H.-E. bundle if it admits an H.-E. connection A for some Hermitian metric h. If this is the case, one says that h is a **Hermitian-Einstein structure** on the holomorphic bundle $\mathcal{E}_A$ defined by A [K2].
The constant c in the definition of H.-E. connections is essentially determined by **cohomological invariants** of (X, g) and E:

$$c = \frac{\pi}{\mathrm{vol}_g(X)} \cdot \frac{\deg_\omega(\mathrm{E})}{\mathrm{r}},$$

where the ω-degree of E is the real number

$$\deg_\omega(\mathrm{E}) = \mathrm{c}_1(\mathrm{E}) \cdot [\omega].$$

Hermitian-Einstein bundles have three basic properties.

i) Lübke's inequality: Let E/X be a complex vector bundle of rank r and P the $U(r)$-bundle corresponding to a Hermitian metric h on E. For any connection $A \in \mathcal{A}(P)$ there exists two decompositions of F_A:

$$F_A = F_A^{2,0} + F_A^{0,2} + \hat{F}_A \cdot \omega + F_{A\perp}^{1,1}$$

$$F_A = \frac{1}{r}\mathrm{tr}\,(F_A) \cdot \mathrm{id}_E + F_A^0.$$

Comparing the components one sees that A is a Hermitian-Einstein connection iff $\mathrm{tr}\,(F_A)$ is harmonic of type (1,1) and F_A^0 an anti-self-dual 2-form.
In other words, H.-E. connections are the absolute minima of the Yang-Mills functional $YM_X^g(P)$. The trace free component F_A^0 is the curvature of the induced $PU\,(r)$- connection $\bar{A}$. Applying the Chern-Weil formula

$$\|F_{\bar{A}-}\|^2 - \|F_{\bar{A}+}\|^2 = [2\,r\,c_2(E) - (r-1)c_1(E)^2] \cdot \frac{4\pi^2}{r}$$

for $\bar{A}$ one gets **Lübke's inequality.**

Theorem 12 [Lü2]. *The Chern classes of a Hermitian-Einstein bundle E/X satisfy*

$$2\,r\,c_2(E) - (r-1)c_1(E)^2 \geq 0$$

with equality iff the projective bundle $\mathbb{P}(E)$ admits a flat $PU\,(r)$-structure.

ii) Kobayashi's vanishing theorem: Let A be a connection on a Hermitian bundle E. Comparing the **Laplacians** Δ_A and $\bar{\Delta}_A$ of d_A and $\bar{\partial}_A$ on $A^0(E)$ one finds the formula [D1]

$$\bar{\Delta}_A = \frac{1}{2}(\Delta_A - i\hat{F}_A).$$

If A is an H.-E. connection with negative constant c, then $\bar{\Delta}_A = \frac{1}{2}(\Delta_A - c\,\mathrm{id}_E)$ is a positive operator. This implies the following result of Kobayashi.

Theorem 13 [K1]. *Suppose that E/X admits a Hermitian-Einstein connection A with constant c. If $c < 0$, then $H^0(X;\mathcal{E}_A) = 0$. If $c = 0$, then every holomorphic section of $\mathcal{E}_A$ is parallel.*

An H.-E. connection on E is said to be **reducible** if its holonomy is contained in a subgroup of type $U(k) \times U(r-k)$ for some $0 < k < r$ [D1]. This means that the bundle splits as an orthogonal sum of Hermitian-Einstein bundles with the same constants. **Irreducible** H.-E. connections A define **simple** holomorphic bundles $\mathcal{E}_A$ since by Kobayashi's vanishing theorem $\mathrm{End}\,\mathcal{E}_A$ has only parallel holomorphic sections.

iii) Stability: The ω-degree of a torsionfree coherent sheaf $\mathcal{F}$ over X is defined as the ω-degree

$$\deg_\omega(\mathcal{F}) = c_1(\det \mathcal{F}) \cdot [\omega]$$

of its determinant line bundle $\det \mathcal{F}$ [K2]. A holomorphic vector bundle $\mathcal{E}$ of rank r over the Kähler surface (X, g) is **ω-stable** (**ω-semi-stable**) if every coherent subsheaf $\mathcal{F} \subset \mathcal{E}$ with $0 < \mathrm{rk}\,\mathcal{F} < r$ satisfies

$$\frac{\deg_\omega(\mathcal{F})}{\mathrm{rk}\,(\mathcal{F})} < \frac{\deg_\omega(\mathcal{E})}{\mathrm{r}} \quad \left(\frac{\deg_\omega(\mathcal{F})}{\mathrm{rk}\,\mathcal{F}} \leq \frac{\deg_\omega(\mathcal{E})}{\mathrm{r}}\right).$$

Theorem 14 [Lü2], [K2]. *Let $\mathcal{E}$ be a holomorphic vector bundle which admits a Hermitian-Einstein structure. Then $\mathcal{E}$ is either ω-stable or a direct sum of ω-stable bundles with the same degree/rank ratio.*

The idea of the proof is very simple [K2]: A coherent subsheaf $\mathcal{F} \subset \mathcal{E}$ of rank s induces a nontrivial morphism $\det \mathcal{F} \to \Lambda^s \mathcal{E}$ which can be interpreted as a non-zero section in the holomorphic bundle $\Lambda^s \mathcal{E} \otimes (\det \mathcal{F})^\vee$. The latter bundle admits a Hermitian-Einstein structure with constant $\frac{\pi}{\mathrm{vol}_g(X)}\left[s \cdot \frac{\deg_\omega(\mathcal{E})}{\mathrm{r}} - \deg_\omega(\mathcal{F})\right]$. By Kobayashi's vanishing theorem this constant must be ≥ 0, i.e. $\mathcal{E}$ is at least ω-semi-stable. The other assertions follow from a closer analysis of the case $s \cdot \frac{\deg_\omega(\mathcal{E})}{\mathrm{r}} = \deg_\omega(\mathcal{F})$ [Lü2].
The converse to this result, which became known as the **Kobayashi-Hitchin conjecture** [Mar], has been proved by Donaldson in the projective-algebraic case [D1] and by Uhlenbeck-Yau in the general Kähler case [UY].

Theorem 15 [D1], [UY]. *Every ω-stable vector bundle over X admits a Hermitian-Einstein structure. The Hermitian-Einstein connection is unique.*

This result is the essential step for the identification of instanton moduli spaces over algebraic surfaces with moduli spaces of stable algebraic vector bundles.

4.2 Stable algebraic bundles

For the rest of this section let X be a smooth complex **projective-algebraic** surface **polarized** by the class of a very ample divisor H. Denote the degree/rank ratio of a non-trivial coherent torsionfree algebraic sheaf $\mathcal{F}$ with respect to H by $\mu(\mathcal{F}) := \frac{c_1(\mathcal{F})H\cdot H}{\mathrm{rk}\,(\mathcal{F})}$. An algebraic vector bundle $\mathcal{E}$ over X is **H-stable** (**H-semi-stable**) if every coherent subsheaf $\mathcal{F} \subset \mathcal{E}$ with $0 < \mathrm{rk}\,(\mathcal{F}) < \mathrm{rk}\,(\mathcal{E})$ satisfies $\mu(\mathcal{F}) < \mu(\mathcal{E})$ ($\mu(\mathcal{F}) \leq \mu(\mathcal{E})$) [OSS].
Let $M_X^H(E)$ be the set of algebraic isomorphism classes $[\mathcal{E}]$ of H-stable algebraic bundles over X which are differentiably equivalent to the fixed C^∞-bundle E [OSS]. A **family** of H-stable bundles in $M_X^H(E)$ parametrized by an algebraic variety S is an algebraic vector bundle $\mathbb{E}$ over $S \times X$, such that $\mathbb{E}_s := \mathbb{E}|_{\{s\}\times X}$ defines an element in $M_X^H(E)$ for every parameter $s \in S$. Denote by $\| \mathbb{E} \|$ the equivalence class of such a family under the equivalence relation $\mathbb{E} \sim \mathbb{E}'$ iff $\mathbb{E}' \cong \mathbb{E} \otimes \mathrm{pr}_S^* \mathcal{L}$ for some algebraic line bundle $\mathcal{L}$ over S. Let

$$M_X^H(E)(-) : \underline{(Alg)} \to \underline{(Set)}$$

be the functor from algebraic varieties to sets which assigns the set $M_X^H(E)(S) := \{\| \mathbb{E} \| \mid \mathbb{E}/S \otimes X \text{ a family with } [\mathbb{E}_s] \in M_X^H(E) \forall s \in S\}$ to a variety S.

An algebraic variety $\mathcal{M}_X^H(E)$ is a **coarse moduli** space for $M_X^H(E)(-)$ if there exists a morphism

$$M_X^H(E)(-) \to \mathrm{Hom}\,(-, \mathcal{M}_X^H(E))$$

which induces a bijection for a point, so that $\mathcal{M}_X^H(E)$ is **minimal** in the following sense: if $\mathcal{N}$ is another variety with the above mentioned properties, then there exists a unique morphism $\mathcal{M}_X^H(E) \xrightarrow{u} \mathcal{N}$ such that the following diagram commutes:

$$\begin{array}{ccc} M_X^H(E)(-) & \longrightarrow & \mathrm{Hom}\,(-, \mathcal{M}_X^H(E)) \\ & \searrow & \swarrow u_* \\ & \mathrm{Hom}\,(-, \mathcal{N}) & \end{array}$$

A coarse moduli space is unique up to isomorphism if it exists.

Theorem 16 [M1], [G]. *Let X be a smooth projective-algebraic surface with a very ample divisor H. There exists a quasi-projective variety $\mathcal{M}_X^H(E)$ which is a coarse moduli space for $M_X^H(E)(-)$.*

Using **geometric invariant theory** the moduli space $\mathcal{M}_X^H(E)$ can be constructed as a Zariski open subset of a geometric quotient $Q^s/_{PGL}$ of stable points in a certain Quot-scheme [M2], [M3].
From this construction it is almost obvious that the germ of the underlying complex analytic space $\mathcal{M}_X^H(E)_{an}$ at any point $[\mathcal{E}]$ is the base space of the universal deformation of the analytic vector bundle $\mathcal{E}_{an}$ [We]. Moreover, $\mathcal{M}_X^H(E)_{an}$ is a coarse moduli space for the obvious functor $M_X^H(E)_{an}(-)$ from complex analytic varieties to sets.

Next we collect the main properties of these moduli spaces.

i) Local structure: By general deformation theory the **Zariski tangent space** of $\mathcal{M}_X^H(E)$ at a point $[\mathcal{E}]$ is the cohomology group $H^1(X;\mathrm{End}\,\mathcal{E})$. The split exact sequence

$$0 \to \mathcal{O}_X \overset{\frac{1}{r}tr}{\underset{\curvearrowleft}{\longrightarrow}} \mathrm{End}\,\mathcal{E} \to \mathrm{End}_0\mathcal{E} \to 0$$

induces a decomposition $H^1(X;\mathrm{End}\,\mathcal{E}) \cong H^1(X;\mathcal{O}_X) \oplus H^1(X;\mathrm{End}_0\,\mathcal{E})$. The cohomology $H^1(X;\mathrm{End}_0\,\mathcal{E})$ of trace free endomorphisms is the Zariski tangent space of the base of the versal deformation of the associated **projective bundle** $\mathbb{P}(\mathcal{E}_{an})$. In fact $\mathcal{M}_X^H(E)_{an}$ is locally a product of this base space and the Picard variety $\mathrm{Pic}^0(X)$ of X [EF]. The base of the versal deformation of $\mathbb{P}(\mathcal{E}_{an})$ can be described as the inverse image of $0 \in H^2(X;\mathrm{End}_0\,\mathcal{E})$ under an analytic **Kuranishi map**

$$K_{[\mathcal{E}]} : H^1(X;\mathrm{End}_0\,\mathcal{E}) \to H^2(X;\mathrm{End}_0\,\mathcal{E}).$$

The dimension of $\mathcal{M}_X^H(E)$ at $[\mathcal{E}]$ is therefore bounded from below:

$$\dim_{[\mathcal{E}]} \mathcal{M}_X^H(E) \geq 2r\,c_2(E) - (r-1)\,c_1\,(E)^2 - (r^2-1)\chi(\mathcal{O}_X) + q(X).$$

If $H^2(X;\mathrm{End}_0\,\mathcal{E})$ vanishes, then $\mathcal{M}_X^H(E)$ is smooth at $[\mathcal{E}]$.
This follows immediately from the Kuranishi description and the Riemann-Roch theorem. The lower bound is the **expected dimension** of $\mathcal{M}_X^H(E)$.

ii) Global structure: A moduli space $\mathcal{M}_X^H(E)$ is said to be a **fine** moduli space if there exists a **universal family** $\mathbb{E}$ parametrized by $\mathcal{M}_X^H(E)$. This means that $\mathbb{E}$ is an algebraic vector bundle over $\mathcal{M}_X^H(E) \times X$, so that its restriction $\mathbb{E}_{[\mathcal{E}]}$ to $\{[\mathcal{E}]\} \times X$ defines the point $[\mathcal{E}]$. In this case $(\mathcal{M}_X^H(E), \mathbb{E})$ represents the functor $M_X^H(E)(-)$.
In general such universal bundles do not exist. There is, however, a useful sufficient numerical condition — due to Maruyama — which guarantees the existence of a universal family.
Write the Hilbert polynomial of $\mathcal{E}$ with respect to the line bundle $\mathcal{O}_X(H)$ in the form

$$\sum_{i=0}^{2}(-1)^i h^i(X;\mathcal{E}\otimes\mathcal{O}_X(mH)) = \sum_{k=0}^{2} a_k \binom{m+k}{k}$$

with integer coefficients a_k.

Theorem 17 [M3]. *A universal bundle* $\mathbb{E}$ *over* $\mathcal{M}_X^H(E) \times X$ *exists if* $g.c.d.(a_0, a_1, a_2) = 1$.

Example 3. Let $\mathcal{E}$ be a rank-2 bundle with Chern classes $c_i = c_i(\mathcal{E})$. Then this criterion reads

$$g.c.d.(2H^2, \chi(\mathcal{E}), H\cdot(H - K + c_1)) = 1.$$

Here K is a canonical divisor, $\cdot$ denotes the intersection product and $\chi(\mathcal{E})$ is the Euler characteristic $\chi(\mathcal{E}) = \sum_{i=0}^{2}(-1)^i h^i(X,\mathcal{E})$.

The natural algebraic structure of the spaces $\mathcal{M}_X^H(E)$ is in general **non-reduced**; there even exist examples which are non-reduced at the generic points of all irreducible components [FM3], [OV1]. Exept for these there not many general results relating to the global structure of the moduli spaces. Very often global properties of $\mathcal{M}_X^H(E)$ reflect global properties of the base [B], [MU].

4.3 Instantons over algebraic surfaces

Fix a smooth projective surface $X \subset \mathbb{P}^N$ and a differentiable complex vector bundle E of rank r over X. Let g be the Hodge metric obtained from the Fubini-Study metric on $\mathbb{P}^N$. The corresponding Kähler form ω represents the dual class of a hyperplane section H, so that ω-stability and H-stability become equivalent concepts.
Let $P = P_h$ be the unitary frame bundle associated to the choice of a Hermitian metric h on E, and denote by

$$\mathcal{A}^{1,1}_{\mathrm{H.-E.}}(P) := \{A \in \mathcal{A}^{1,1}(P) \mid i\hat{F}_A = c \cdot \mathrm{id}_E\}$$

the set of Hermitian-Einstein connections on P. The gauge group $\mathrm{Aut}(P)$ acts on $\mathcal{A}^{1,1}_{\mathrm{H.-E.}}(P)$ leaving the subset $\mathcal{A}^{1,1}_{\mathrm{H.-E.}}(P)^*$ of irreducible H.-E. connections invariant. The assignment $A \mapsto \bar{\partial}_A$ induces a well-defined map

$$\Phi_h : \mathcal{A}^{1,1}_{\mathrm{H.-E.}}(P_h)^* / \mathrm{Aut}(P_h) \to \mathcal{M}^s_X(E)$$

into the moduli space of simple holomorphic structures. These maps Φ_h are **open embeddings** whose images are independent of h [Ki], [LO], [K2]. Denote the common image by $\mathcal{M}_{\mathrm{H.-E.}}(E)^*$; this is the moduli space of **irreducible Hermitian-Einstein structures** on E.

Theorem 18 [D1], [FuS], [K2], [LO], [Mi]. *The moduli space $\mathcal{M}_{\mathrm{H.-E.}}(E)^*$ of irreducible Hermitian-Einstein structures on E is a complex space. It is naturally isomorphic to the underlying complex space $\mathcal{M}_X^H(E)_{\mathrm{an}}$ of the moduli space of stable algebraic vector bundles.*

The stability of irreducible H.-E. structures and the solution of the Kobayashi-Hitchin conjecture mentioned above show immediately that
$\mathcal{M}_{\mathrm{H.-E.}}(E)^* \cong \mathcal{M}_X^H(E)$ as sets. In order to obtain the equivalence of both spaces as complex spaces one has to invoke a result due to Miyajima [Mi]. He shows in particular that $\mathcal{M}^s_X(E)$ is the complex space underlying the moduli space of **simple algebraic** bundles. The latter has been constructed by Altman and Kleiman [AK].
As a consequence of the last theorem we have the following algebraic interpretation of instantons over algebraic surfaces.

Theorem 19 [D4], [OV2]. *Let E/X be a differentiable vector of rank 2 over a simply-connected smooth projective surface $X \subset \mathbb{P}^N$. Denote by g and H the induced Hodge metric and a hyperplane section respectively. If $c_1(E) \in NS(X)$, then every choice of a unitary structure P on E induces an isomorphism*

$$\mathcal{M}_X^g(P/_{S^1})^* \cong \mathcal{M}_X^H(E)$$

of the underlying differentiable spaces.

To see this one associates to an H.-E. connection $A \in \mathcal{A}^{1,1}_{\mathrm{H.-E.}}(P)$, whose central curvature component $\frac{1}{2}\,\mathrm{tr}\,(F_A)$ is equal to the g-harmonic representative of $-\pi i\, c_1(E)$, the induced instanton $\bar{A}$ on the associated $PU(2)$-bundle $P/_{S^1}$. This identifies $\mathcal{M}_X^g(P/_{S^1})^*$ with the moduli space $\mathcal{A}^{1,1}_{\mathrm{H.-E.}}(P)^*/_{\mathrm{Aut}(P)}$ [OV2].

If $\mathcal{M}_X^H(E)$ is smooth, then its **complex orientation** induces the natural orientation on the corresponding instanton space [D5]. Furthermore, if $\mathcal{M}_X^H(E)$ happens to be a **fine** moduli space space with a universal family, $\mathbb{E}$, then the projectivized bundle $\mathbb{P}(\mathbb{E})$ is equivalent to the universal bundle $\bar{\mathbb{P}}$ over $\mathcal{M}_X^g(P/_{S^1})^* \times X$.

The choice of a $SU(2)$-structure P on bundles with $c_1(E) = 0$ yields an identification of $\mathcal{M}_E^H(X)$ with the moduli space of $SU(2)$-instantons on P [D4].

5 C^∞-structures of algebraic surfaces

This section contains applications of the Donaldson invariants to the differential topology of algebraic surfaces. In this situation algebro-geometric techniques can be used to show that the invariants are non-trivial in general, and to calculate them explicitly in good cases.

5.1 Topology of algebraic surfaces

Let X be a closed oriented simply-connected 4-manifold with **intersection form**

$$S_X : H_2(X;\mathbb{Z}) \times H_2(X;\mathbb{Z}) \to \mathbb{Z}.$$

This is a symmetric bilinear form which is unimodular by Poincaré duality.
Quite generally one says that a symmetric bilinear form $S : L \times L \to \mathbb{Z}$ is **even** (of type II) if its associated quadratic form $q : L \to \mathbb{Z}$ attains only even values; otherwise S is said to be **odd** (of type I). The form S is positive (negative) **definite** if $q > 0$ ($q < 0$), **indefinite** if both signs occur.
Indefinite unimodular forms are up to equivalence over $\mathbb{Z}$ characterized by three invariants: the **rank**, the **signature**, and the **type** [S]. If $S = S_X$ is the intersection form of a 4-manifold X as above, then these invariants are simply $b_2(X), b_+(X) - b_-(X)$ and $w_2(X)$ (being zero or not). Freedman's fundamental classification theorem then takes the following form.

Theorem 20 [Fr]. *The natural map from oriented homeomorphism types of simply-connected topological 4-manifolds to $\mathbb{Z}$-equivalence classes of unimodular symmetric bilinear forms, given by $X \mapsto S_X$, is 1-1 for even forms and 2-1 for odd forms.*

The two distinct topological realizations X, X' of a given odd form S are distinguished by the fact that exactly one of the 5-manifolds $X \times S^1$ and $X' \times S^1$ is smoothable.

Intersection forms of algebraic surfaces other than the projective plane $\mathbb{P}^2$ are always **indefinite** [Y]. Freedman's theorem and the classification of indefinite forms has therefore a simple corollary.

Theorem 21. *The homeomorphism type of a simply-connected algebraic surface X is determined by its Chern numbers c_1^2, c_2 and the parity of its canonical class k_X.*

Example 4. Using standard notation [BPV] one has $S_{\mathbb{P}^2} =< 1 >$, $S_{\mathbb{P}^1 \times \mathbb{P}^1} =< H >$, and $S_K = -2 < E_8 > \oplus 3 < H >$ for the intersection forms of the plane, the quadric and of a $K3$ surface K.

Every other simply-connected algebraic surface X is **homeomorphic** to a connected sum of these three surfaces eventually with the opposite orientation.

Example 5. Let $X^k \subset \mathbb{P}^1 \times \mathbb{P}^2$ be a smooth hypersurface defined by a bihomogeneous polynomial of bidegree $(k+1, 3)$. Such a surface is simply-connected by the Lefschetz theorem, and the projection $\mathbb{P}^1 \times \mathbb{P}^2 \to \mathbb{P}^1$ induces on X^k the structure of a relatively minimal **elliptic surface** over $\mathbb{P}^1$. Perform **logarithmic transformations** of multiplicities p and q along two smooth fibers [BPV]. If p and q are relatively prime, the resulting surface $X^k_{p,q}$ is again simply-connected and relatively minimal elliptic over $\mathbb{P}^1$ [BPV], [Ue]. Denote a generic fiber of $X^k_{p,q}/\mathbb{P}^1$ by F and let $p\,F_p \sim q\,F_q \sim F$ be the multiple fibers. The **canonical bundle formula** for $X^k_{p,q}$ yields [BPV]:

$$K_{X^k_{p,q}} \sim (k-1)F + (p-1)F_p + (q-1)F_q.$$

It implies $p_g(X^k_{p,q}) = k$ and shows that the **divisibility** of the canonical class of $X^k_{p,q}$ is equal to $(k+1)pq - (p+q)$.
The intersection form of $X^k_{p,q}$ is therefore of type II if k, p, and q are odd and otherwise of type I.
Since by **Noether's formula** $c_2(X^k_{p,q}) = 12(k+1)$, we find ($\approx$ means homeomorphic):

$$X^k_{p,q} \approx \begin{cases} \#_{\frac{k+1}{2}} K \; \# \; \#_{\frac{k-1}{2}} \mathbb{P}^1 \times \mathbb{P}^1 & \text{if } k \equiv p \equiv q \equiv 1 \pmod 2 \\ \#_{2k+1} \mathbb{P}^2 \; \# \; \#_{10k+9} \bar{\mathbb{P}}^2 & \text{otherwise.} \end{cases}$$

Here $\bar{\mathbb{P}}^2$ denotes $\mathbb{P}^2$ with the opposite orientation.

Note that $X^0_{p,q}$ is **rational** if $\min(p,q) = 1$; the surfaces $X^1_{1,1}$ are $K3$ **surfaces**, but all the other $X^k_{p,q}$ are of Kodaira dimension 1 [BPV]. The surfaces $X^0_{p,q}$ with $\min(p,q) > 1$ are named after **Dolgachev** who constructed them as counter-examples to the Severi problem [Dol].
The surfaces $X^1_{p,q}$ with $p \equiv q \equiv 1 \pmod 2$ and $p \cdot q \neq 1$ are honestly elliptic surfaces but homeomorphic to a $K3$ surface; they are examples of so-called **homotopy** $K3$ **surfaces** [Ko].

The **Enriques-Kodaira classification** allows to classify simply-connected algebraic surfaces up to **deformation** [BPV], [FM4].

Theorem 22. *Every simply-connected algebraic surface which is not of general type is a deformation of one of the following surfaces:*

i) $\hat{\mathbb{P}}^2(z_1, \ldots, z_n), n \geq 0$.
ii) $\mathbb{P}^1 \times \mathbb{P}^1$.
iii) $\hat{X}^k_{p,q}(x_1, \ldots, x_m), m \geq 0$, *with* $\min(p,q) > 1$ *if* $k = 0$.

There exist only a finite number of deformation types of surfaces of Kodaira dimension 2 in a given homotopy type.

Here $\hat{X}(x_1, \ldots, x_m)$ denotes the blow up of a surface X in m distinct points.

Example 6 [Ba]. Let $Q \subset \mathbb{P}^4$ be the complete intersection with equations

$$\sum_{i=0}^{4} Z_i = 0 \quad , \quad \sum_{i=0}^{4} Z_i^5 - \frac{5}{4}\left(\sum_{i=0}^{4} Z_i^3\right)\left(\sum_{i=0}^{4} Z_i^2\right) = 0.$$

This quintic has 20 nodes — the points in the orbit of $< 2, 2, 2, -3-\sqrt{-7}, 3+\sqrt{7} >$ under the obvious action of the alternating group A_5 — and no other singularities. It is the canonical image of a smooth surface $\hat{Q}$ with the 1-canonical map $\varphi_{|K_{\hat{Q}}|} : \hat{Q} \to Q$ being the orbit map of a canonical involution $\imath : \hat{Q} \to \hat{Q}$. From this involution and the A_5-action on Q it is possible to construct an action of the dihedral group D_{10} on $\hat{Q}$, so that the quotient $\hat{Q}/_{D_{10}}$ is smooth except for 4 nodes $p_1, \ldots, p_4$. The minimal resolution $\tau : Y \to \hat{Q}/_{D_{10}}$ of these singularities yields a smooth minimal surface Y of general type which is simply-connected with Chern numbers $c_1^2 = 1, c_2 = 11$. This **Barlow surface** Y is therefore homeomorphic to the projective plane $\hat{\mathbb{P}}^2(z_1, \ldots, z_8)$ blown up in 8 points; it is at present the only confirmed example (up to deformations) of a minimal surface of general type which is homeomorphic to a rational surface.

We will see that Y is **not diffeomorphic** to $\hat{\mathbb{P}}^2(z_1, \ldots, z_8)$.

5.2 Exotic C^∞-structures on $\mathbb{P}^2 \# 9\,\overline{\mathbb{P}}^2$

The topological 4-manifold $\mathbb{P}^2\#9\overline{\mathbb{P}}^2$ supports infinitely many distinct algebraic structures: It is the underlying manifold of $\hat{\mathbb{P}}^2(z_1, \ldots, z_9)$, the Dolgachev surfaces $X^0_{p,q}$ for all sets $\{p, q\}$ of coprime multiplicities p, q with $\min(p, q) > 1$, and of the blow-up $\hat{Y}(y)$ of Barlow's surface. These algebraic surfaces can be distinguished by the behaviour of their pluricanonical linear systems [BPV].
In his seminal paper [D3] Donaldson shows that the underlying smooth structure of the Dolgachev surface $X^0_{2,3}$ is distinct from the standard structure, i.e. the one induced by $\hat{\mathbb{P}}^2(z_1, \ldots, z_9)$. He uses his Γ-invariant to distinguish between the two diffeomorphism types.
His theorem has later been extended by R. Friedman and J. Morgan [FM2], [FM3], and by A. Van de Ven and the author of this article [OV1] to include also the other Dolgachev surfaces. The result is that the surfaces $X^0_{p,q}$ define infinitely many different C^∞-structures on the topological manifold $\mathbb{P}^2\#9\overline{\mathbb{P}}^2$, showing that the smooth h-cobordism conjecture for 4-manifolds fails in a very essential way.
The use of the $\bar{\Gamma}$-invariant instead of the Γ-invariant led to a simplification of the proof and allowed at the same time also to treat the Barlow surface.

Theorem 23 [OV2]. *The Barlow surface $\hat{Y}(y)$ blown up in one point is neither diffeomorphic to $\hat{\mathbb{P}}^2(z_1, \ldots, z_9)$ nor to any Dolgachev surface.*

Our result implies of course that the Barlow surface itself is not diffeomorphic to $\hat{\mathbb{P}}^2(z_1, \ldots, z_8)$ [Kot].

The proof of the theorem goes as follows [OV2]: Let X be one of the surfaces $\hat{\mathbb{P}}^2(z_1, \ldots, z_9), X^0_{p,q}$ with $\min(p, q) > 1$, or $\hat{Y}(y)$.
Over X there is a unique irreducible $PU(2)$-bundle P with $w_2(P) = w_2(X)$ and $p_1(P) = -4$. Consider the $\bar{\Gamma}$-invariant corresponding to this bundle and the canonical lift $k_X = c_1(K_X)$ of $w_2(X)$:

$$\bar{\Gamma}_{X,k_X} : \{\Omega_+, \Omega_-\} \to H^2(X; \mathbb{Z}).$$

Choose a suitable embedding $X \subset \mathbb{P}^N$ of each surface, and denote by g and H the induced Hodge metric and a corresponding hyperplane section respectively. The $PU(2)$-instanton space $\mathcal{M}^g_X(P)^*$ can then be identified with the moduli space $\mathcal{M}^H_X(k_X, 1)$ of H-stable algebraic vector bundles $\mathcal{E}$ of rank 2 with Chern classes $c_1(\mathcal{E}) = k_X, c_2(\mathcal{E}) = 1$. It turns out that $\mathcal{M}^H_X(k_X, 1)$ is a (possibly empty) disjoint union of finitely many complete curves $\mathcal{M}_i$ whose natural scheme structure is in general not reduced [OV2]. Using Maruyama's criterion one finds a universal bundle $\mathbb{E}$ over $\mathcal{M}^H_X(k_X, 1) \times X$, so that the value of $\bar{\Gamma}_{X,k_X}$ on the chamber Ω_+ which contains the class $[H]$ is given by

$$\bar{\Gamma}_{X,k_X}(\Omega_+) = \sum_i m_i p_1(\mathbb{E}_i)/[\hat{\mathcal{M}}_i].$$

Here $\hat{\mathcal{M}}_i$ denotes the normalization of the reduction of the i-th curve $\mathcal{M}_i, m_i$ the multiplicity of $\mathcal{M}_i$, and $\mathbb{E}_i$ the pullback of $\mathbb{E}$ to $\hat{\mathcal{M}}_i$ [OV2].
To determine the moduli spaces $\mathcal{M}_X^H(k_X, 1)$ and the universal bundles $\mathbb{E}$ a technique is used which is known as **Serre's construction** [OSS]. This construction relates vector bundles over a surface to finite sets of points on the surface which are in special position with respect to certain linear systems [GH], [OV3]. In the situation at hand the 2-canonical system $|2K_X|$ is the relevant linear system. Its base locus $Bs|2K_X|$ determines the moduli space $\mathcal{M}_X^H(k_X, 1)$. We summarize the result of the computations in the following table:

X	kod(X)	$P_2(X)$	$Bs\|2K_X\|$	$\widehat{M_X^H(k_X,1)}$	$\bar{\Gamma}_{X,k_X}(\Omega_+)$
$\hat{\mathbb{P}}^2(z_1,\ldots,z_9)$	$-\infty$	0	$\emptyset$	$\emptyset$	0
$X^0_{p,q}$	1	1	$(p-2)F_p+(q-2)F_q$	$\coprod_{\alpha_{p,q}} F_p \amalg \coprod_{\beta_{p,q}} F_q$	$\gamma_{p,q}\, \kappa_{X^0_{p,q}}$
$\hat{Y}(y)$	2	2	$\{y_1,\ldots,y_4\} \amalg E$	$\coprod_{i=1}^{4}\{z_i\} \times E$	$-4m[E]$

Here kod(X) and $P_2(X)$ are the Kodaira dimension and the 2-genus of X respectively. F_p and F_q denote the reductions of the multiple fibers of $X^0_{p,q}$, $\kappa_{X^0_{p,q}}$ is a primitive class with $q[F_p] = p[F_q] = p\,q\,\kappa_{X^0_{p,q}}$; $\alpha_{p,q}, \beta_{p,q}$, and $\gamma_{p,q}$ are natural numbers with $\gamma_{p,q} \geq q\,\alpha_{p,q} + p\,\beta_{p,q}$. The four points $\{y_1,\ldots,y_4\} \subset Y$ form the base locus of $|2K_Y|, E \subset \hat{Y}(y)$ is the exceptional curve, m a positive integer [OV2].
It is obvious from this table that the $\bar{\Gamma}$-invariant distinguishes the C^∞-structures of the surfaces with distinct Kodaira dimension in the table; furthermore, the inequality $\gamma_{p,q} \geq q\,\alpha_{p,q} + p\,\beta_{p,q}$ shows the existence of infinitely many different smooth structures coming from Dolgachev surfaces.
Note that the 2-canonical system whose properties determine the value of the $\bar{\Gamma}$-invariant is the same linear system which — by **Castelnuovo's criterion** — determines whether or not a simply-connected surface is rational [BPV].

5.3 Indecomposability of algebraic surfaces

One of the main unsolved problems in 4-dimensional topology is the following question which goes back to R. Thom: Are all simply-connected 4-manifolds connected sums of algebraic surfaces?
A related question is whether algebraic surfaces themselves can be decomposed into smaller pieces.

Theorem 24 [D4]. *If a simply-connected algebraic surface X decomposes as a smooth connected sum $X \cong X_1 \# X_2$, then one of the 4-manifolds X_i has a negative definite intersection form.*

A m-fold blow up of an algebraic surface X realizes such a decomposition $\hat{X}(x_1, \ldots, x_m) \cong X \# m\bar{\mathbb{P}}^2$ with a negative definite piece $m\bar{\mathbb{P}}^2$. Donaldson conjectures that this is the only possibility, i.e. minimal models of simply-connected algebraic surfaces are indecomposable as differentiable 4-manifolds [D4]. At least for surfaces with Kodaira dimension ≥ 0 this is very likely.
The following corollary is essentially a consequence of Donaldson's indecomposability theorem.

Corollary 25 [D7]. *Every simply-connected algebraic surface with $p_g > 0$ admits an exotic C^∞-structure.*

Indeed, given such a surface X, there exists a homeomorphism

$$X \approx \begin{cases} a(\pm K) & \# \quad b\,\mathbb{P}^1 \times \mathbb{P}^1 \quad \text{if} \quad S_X \quad \text{is even} \\ m\mathbb{P}^2 & \# \quad n\,\bar{\mathbb{P}}^2 \quad \text{if} \quad S_X \quad \text{is odd.} \end{cases}$$

Here $\pm K$ denotes a $K3$ surface with one of the possible orientations.
The decomposable manifold on the right defines an exotic structure unless $a = 1$ and $b = 0$, i.e. unless X is a homotopy $K3$ surface. In this case, one needs a result of Friedman and Morgan, saying that homotopy $K3$ surfaces $X^1_{p,q}$ with $p \cdot q \neq 1$ are not diffeomorphic to honest $K3$ surfaces [FM4].

Donaldson's indecomposability theorem is an immediate consequence of his vanishing theorem for the polynomial invariants of connected sums and the following **positivity theorem.**

Theorem 26 [D4]. *Let X be a simply-connected algebraic surface with $p_g(X) > 0$. If H is a hyperplane section, then $\gamma^k_{X,o_+}([H], \ldots, [H]) > 0$ for all sufficiently large k.*

Here $\gamma^k_{X,o_+} : S^{d(k)} H_2(X; \mathbb{Z}) \to \mathbb{Z}$ denotes the Donaldson polynomial associated to a principal $SU(2)$-bundle P_k with second Chern class k.

The proof of the positivity theorem uses the interpretation of instantons over algebraic surfaces as stable vector bundles: Consider an embedding $X \subset \mathbb{P}^N$ with hyperplane section H and associated Hodge metric g. The instanton moduli space $\mathcal{M}^g_X(P_k)^*$ is then diffeomorphic to the moduli space $\mathcal{M}^H_X(k)$ of H-stable algebraic vector bundles $\mathcal{E}$ over X with $c_1(\mathcal{E}) = 0$ and $c_2(\mathcal{E}) = k$.
Donaldson proves that the restriction of the class $\mu([H]) \in H^2(\mathcal{B}^*(P_k); \mathbb{Z})$ to $\mathcal{M}^g_X(P_k)^*$ defines a polarization of $\mathcal{M}^H_X(k)$ via the identification $\mathcal{M}^g_X(P_k)^* \cong \mathcal{M}^H_X(k)$ [D4]. Since $\mathcal{M}^H_X(k)$ is quasi-projective and non-empty for large k [M1], it must have positive degree with respect to this polarization. Donaldson shows that computing this degree is an admissible way of calculating the value $\gamma^k_{X,o_+}([H], \ldots, [H])$ if k is sufficiently large [D4]. This is not clear since the

Hodge metric is not one of the generic metrics which were used in the definition of γ^k_{X,o_+}. The crucial point is to estimate the dimensions of the subspaces $\sum_k \subset \mathcal{M}^H_X(k)$ consisting of bundles $\mathcal{E}$ with $h^2(X;\mathrm{End}_0\,\mathcal{E}) \neq 0$. Donaldson's estimate has recently been generalized by K. Zuo to vector bundles with $c_1 \neq 0$.

Theorem 27 [D4], [Z]. *Let $X \subset \mathbb{P}^N$ be an algebraic surface with hyperplane section H and canonical divisor K. Fix a line bundle $\mathcal{L} \in \mathrm{Pic}(X)$ and let $\mathcal{M}^H_X(\mathcal{L},k)$ be the moduli space of H-stable 2-bundles $\mathcal{E}$ over X with $\det\mathcal{E} \cong \mathcal{L}$ and $c_2(\mathcal{E}) = k$. The subvariety $\sum_k := \{-\mathcal{E}- \in \mathcal{M}^H_X(\mathcal{L},k) | h^2(X;\mathrm{End}_0\,\mathcal{E}) \neq 0\}$ has dimension $\leq 3k + A\sqrt{k} + B$ with constants A, B depending only on $X, \mathcal{L}$ and $[H]$.*

If $\mathcal{M}^H_X(\mathcal{L},k)$ is non-empty, then its dimension is $\geq 4k+C$ with a constant C depending on X and $\mathcal{L}$. If furthermore $4k+C > 3k+A\sqrt{k}+B$, then $\mathcal{M}^H_X(\mathcal{L},k)\setminus\sum_k$ is non-empty of the expected dimension.
Choose k_0 so large that this holds, and let D be the maximum dimension of the moduli spaces $\mathcal{M}^H_X(\mathcal{L},\ell), 1 \leq \ell \leq k_0-1$. If $k \geq k_0$ and $4k+C \geq D+2(k-k_0+1)$, then the degree of $\mathcal{M}^H_X(\mathcal{L},k)$ with respect to the polarization defined by $\mu([H])$, gives the value of the corresponding polynomial invariant on the hyperplane class [D4].

Note that Zuo's generalization of Donaldson's estimate leads to a positivity theorem for the polynomial invariants corresponding to all bundles with $c_1 \in NS(X)$ over simply-connected algebraic surfaces.

5.4 Diffeomorphism groups

Let X be a simply-connected smooth algebraic surface. The component group $\mathrm{Diff}_+(X)$ of the group of orientation preserving diffeomorphisms has a natural representation $\psi : \mathrm{Diff}_+(X) \to O(L)$ in the orthogonal group of the lattice $L := H_2(X;\mathbb{Z})$. The group $O(L)$ can be identified with the group of **isotopy** classes of **self-homeomorphisms** of X; the image of ψ, on the other hand, is isomorphic to the group of **pseudo-isotopy** classes of **diffeomorphisms** and is, in general, difficult to compute [Q]. We will describe it for certain types of algebraic surfaces.

i) Polynomial invariants of reflection groups: Let V be a complex vector space of dimension n with a non-degenerate symmetric bilinear form $<,>: V \times V \to \mathbb{C}$. Denote by $q : V \to \mathbb{C}$ the associated quadratic form, and let $O(V)$ be the corresponding orthogonal group. For a non-isotropic vector $v \in V$ let

$$s_v(x) := x - 2\,\frac{<x,v>}{<v,v>}\,v$$

be the reflection in the hyperplane orthogonal to v. For any subset $\Delta \subset V$ of non-isotropic vectors define $\Gamma_\Delta \subset O(V)$ as the subgroup generated by the reflections s_δ, $\delta \in \Delta$.

Let $k : V \to \mathbb{C}$ be a linear form on V, set $V' := \operatorname{Ker} k$, and let $O_k(V)$ be the subgroup of $O(V)$ consisting of automorphisms preserving k. Write $S = \operatorname{Sym}(V^*)$ for the algebra of polynomial functions on V. For any subgroup $G \subset GL(V)$ let $\bar{G}$ be its Zariski closure and denote by $S^G = S^{\bar{G}}$ the subalgebra of G-invariant polynomials in S.
We want to determine S^{Γ_Δ} for suitable subsets $\Delta \subset V'$.

Theorem 28 **[EO1].** *Suppose $\Delta \subset V'$ is an infinite subset satisfying the conditions*

(A) $< \delta, \delta >= 2$ for all $\delta \in \Delta$.
(B) Δ generates V'.
(C) Δ is a Γ_Δ-orbit.

Then $\bar{\Gamma}_\Delta = O_k(V)$ and $S^{\Gamma_\Delta} = \mathbb{C}[q, k]$.

Here $\mathbb{C}[q, k]$ is the algebra of complex polynomials in q and k; the case $k = 0$ is not excluded.

ii) Surfaces with big diffeomorphism groups: Let X be a simply-connected smooth algebraic surface with intersection form S_X and canonical class $k_X : H_2(X;\mathbb{Z}) \to \mathbb{Z}$. Denote by $q_X : H_2(X;\mathbb{Z}) \to \mathbb{Z}$ the quadratic form associated to S_X and let $O(L)$ be the orthogonal group of the corresponding lattice $L := H_2(X;\mathbb{Z})$.
Set $L' := \operatorname{Ker} k_X, V := H_2(X;\mathbb{C}), q := -q_X$, and $k := k_X$ and define $SO_k(L) \subset O_k(L)$ and $SO_k(V) \subset O_k(V)$ in the obvious way.
A surface X has a **big diffeomorphism group** if $\psi(\mathrm{Diff}_+(X)) \cap SO_k(L)$ is Zariski dense in $SO_k(V)$.

Theorem 29 **[EO1], [EO2], [FMM].** *Let X be a simply-connected smooth algebraic surface. Suppose there exists an infinite subset $\Delta \subset L'$ satisfying the conditions (A), (B) and (C). If every element of Δ is representable by a smooth 2-sphere, then X has a big diffeomorphism group. If $p_g(X) > 0$, then all $SU(2)$-polynomial invariants and all $PU(2)$-invariants associated to bundles with $w_2 = w_2(X)$ are complex polynomials in q_X and k_X.*

The first statement follows since the reflections corresponding to smoothly embedded 2-spheres with self-intersection number -2 are induced by orientation preserving diffeomorphisms [FM2]. The second part follows from the invariance properties of the Donaldson polynomials.

iii) Construction of diffeomorphisms: The usual way to construct subsets Δ satisfying the conditions mentioned above is as sets of **vanishing cycles** of appropriate families [EO1], [FM2]. Let $\pi : \mathcal{X} \to T$ be a smooth proper holomorphic map of connected complex spaces with $X = \pi^{-1}(t_0)$ for some point $t_0 \in T$. The **monodromy group** of this smooth family of surfaces is the subgroup $\Gamma := \operatorname{Im}(\psi \circ \rho) \subset O_k(L)$ where $\rho : \pi_1(T; t_0) \to \mathrm{Diff}_+(X)$ is the monodromy

representation associated to π.
Let $O'_k(L) \subset O_k(L)$ be the kernel of the real spinor norm [EO2]; it consists of those automorphisms in $O_k(L)$ which preserve the components of the subset of maximal positive definite subspaces $H_+ \subset L_{\mathbb{R}}$ in the Grassmannian $\tilde{Gr}_{b_+(X)}(L_{\mathbb{R}})$ of oriented subspaces in $L_{\mathbb{R}}$.

Theorem 30 [EO2]. *Let X be a simply-connected smooth algebraic surface with $p_g > 0$. If X is a complete intersection, a Moishezon or Salvetti surface, then there exists a smooth family $\pi : \mathcal{X} \to T$ containing X with monodromy group $\Gamma = O'_k(L)$.*

Applying complex conjugation to the coefficients of suitable defining polynomial equations for these two types of surfaces one constructs a self-diffeomorphism whose induced map $\sigma : H_2(X;\mathbb{Z}) \to H_2(X;\mathbb{Z})$ sends k_X to $-k_X$ [FM2].

Theorem 31 [EO2]. *Let X be a simply-connected algebraic surface with $p_g \equiv 1$ (mod 2). If X is either a complete intersection with $k_X^2 \equiv 1$ (mod 2), or a Moishezon or Salvetti surface with $k_X^2 \equiv 1$ (mod 2) , then*

$$\psi(\mathrm{Diff}_+(X)) = O'_k(L) \propto \{\sigma, \mathrm{id}\}.$$

The proof of this result has three essential parts: The previous theorem shows that $O'_k(L) \propto \{\sigma, \mathrm{id}\}$ is contained in the image of ψ.
Using the non-triviality and the invariance properties of the $PU(2)$-invariants associated to bundles with $w_2 = w_2(X)$ one shows, quite generally, that $-\mathrm{id} \in O(L)$ is not realizable by an orientation preserving diffeomorphism of a simply-connected algebraic surface with $p_g \equiv 1$ (mod 2) [EO2].
Finally one proves that $\{\pm k_X\}$ is invariant under the image of ψ if k_X divides a non-trivial polynomial invariant of a surface with a big diffeomorphism group.

Example 7. Consider a complete intersection X of multidegree $\underline{d} = (d_1, \ldots, d_r)$ satisfying the following conditions:

i) $d_i \equiv 1$ (mod 2) for $i = 1, \ldots, r$.
ii) $d_j \equiv 0$ (mod 3) for some j.
iii) $\sum_{i<j} e_i \cdot e_j \equiv 1$ (mod 2) where $d_i = 2e_i + 1$ for $i = 1, \ldots, r$.

Then $p_g \equiv 1$ (mod 2) and $k_X^2 \equiv 1$ (mod 2).
An explicit series of examples is given by the multidegrees (3,3), (3,7), (3,11),... .

6 Floer homology

In this section we give a brief description of Floer's instanton homology for 3-dimensional $\mathbb{Z}$-homology spheres and sketch a way for computing these invariants in the Seifert fibered case.

6.1 Instanton homology

Let Σ be a closed oriented 3-dimensional manifold; like every topological 3-manifold Σ carries a unique differentiable structure [He]. Σ is called a **homology sphere** if $H_1(\Sigma;\mathbb{Z}) = 0$, i.e. if Σ has the integral homology of S^3. A well-known example is the Poincaré sphere $\Sigma(2,3,5)$ which can be defined as the quotient of $SU(2)$ by the binary icosahedral group [Br].
Let P be the trivial $SU(2)$-bundle $P = \Sigma \times SU(2)$ over Σ (every $SU(2)$-bundle over a 3-manifold is trivial) and consider the space $\mathcal{B}(P) = \mathcal{A}(P)/_{\mathrm{Aut}(P)}$ of gauge equivalence classes of $SU(2)$-connections.
Using the product connection Θ as base point $\mathcal{A}(P)$ can be identified with the affine space $\mathcal{A}(P) = \Theta + A^1(ad(P))$ where $ad(P) = \Sigma \times LSU(2)$ is the (trivial) adjoint bundle.
The gauge group $\mathrm{Aut}(P)$ is isomorphic to the group $C^\infty(\Sigma, SU(2))$ of C^∞-mappings into $SU(2)$; the assignment $f \mapsto \deg(f)$ identifies its group of components $\pi_0(\mathrm{Aut}(P))$ with the integers. The group $\mathrm{Aut}^*(P) = \mathrm{Aut}(P)/_{\{\pm \mathrm{id}\}}$ acts freely on the open and dense subspace $\mathcal{A}^*(P)$ of irreducible connections, so that the quotient $\mathcal{B}^*(P) = \mathcal{A}^*(P)/_{\mathrm{Aut}(P)}$, or rather a suitable Sobolev completion, becomes an infinite dimensional manifold [F]. The **tangent space** of $\mathcal{B}^*(P)$ at a class $[A]$ is the cokernel of the differential operator $d_A : A^0(ad(P)) \to A^1(ad(P))$, the **cotangent space** $T^*_{\mathcal{B}^*(P)}([A])$ is given by

$$T^*_{\mathcal{B}^*(P)}([A]) = \mathrm{Ker}\,(d_A : A^2(ad(P)) \to A^3(ad(P))).$$

As the curvature $F_A \in A^2(ad(P))$ of A lies in this kernel, the assignment $A \mapsto F_A$ defines a natural 1-form on $\mathcal{B}^*(P)$. This 1-form is locally the differential of the **Chern-Simons function**

$$f : \mathcal{B}(P) \to \mathbb{R}/_{4\mathbb{Z}}$$

which can be defined as follows: consider the path $A_t := (1-t)\Theta + t\,A$ of connections on P as a connection on the bundle $P \times [0,1]$ over $\Sigma \times [0,1]$. Then define

$$f([A]) = \frac{-1}{4\pi^2} \int_{\Sigma \times [0,1]} \mathrm{tr}\,(F_{A_t} \wedge F_{A_t}).$$

The critical set of f, i.e. the zeros of F, is the subset of **flat** connections; via the **holonomy representation** it can be identified with the space

$$R(\Sigma) := \mathrm{Hom}(\pi_1(\Sigma), SU(2))/_{\mathrm{conj.}}$$

of conjugacy classes of **representations** of $\pi_1(\Sigma)$. In general $R(\Sigma)$ is a compact space with components of various dimensions. The subset $R^*(\Sigma) = R(\Sigma)\backslash\{[\Theta]\}$ is the space of **irreducible** connections (since $H_1(\Sigma;\mathbb{Z}) = 0$). We denote by $\alpha = [A]$ the (class of a) representation corresponding to a flat connection A.

Now choose a Riemannian metric σ on Σ with associated Hodge operator $* = *_\sigma$. It induces a Riemannian metric on the moduli space $\mathcal{B}^*(P)$ and allows to identify $T_{\mathcal{B}^*(P)}([A])$ with the kernel

$$T_{\mathcal{B}^*(P)}([A]) = \mathrm{Ker}(d_A^* : A^1(ad(P)) \to A^0(ad(P)))$$

of the adjoint operator d_A^*. Moreover, the 1-form F on $\mathcal{B}^*(P)$ can be converted into a vector field grad $_\sigma(f)$ by setting grad $_\sigma(f)([A]) := *F_A$.
The **gradient lines** of this vector field are families A_t of connections on P satisfying

$$\frac{d}{dt}A_t = - * F_{A_t}.$$

Identifying families of connections on P with connections over the infinite cylinder $\Sigma \times \mathbb{R}$ (with the product metric $\sigma \times 1$) this equation becomes the **anti-self-duality** equation

$$F_{A_t} = - *_{\sigma \times 1} F_{A_t}.$$

In other words, the **trajectories** of the vector field grad$_\sigma(f)$ are **instantons** on the 4-manifold $(\Sigma \times \mathbb{R}, \sigma \times 1)$ [F].

Let $\mathcal{M}_\Sigma^\sigma$ be the moduli space of **finite action instantons** on $\Sigma \times \mathbb{R}$, i.e. instantons A with finite Yang-Mills action $\|F_A\|^2 < \infty$. This moduli space decomposes into spaces $\mathcal{M}_\Sigma^\sigma(\alpha, \beta)$ of instantons approaching classes α, β of flat connections at the ends. Equivalently, $\mathcal{M}_\Sigma^\sigma(\alpha, \beta)$ is the space of **trajectories** of the vector field grad $_\sigma(f)$ connecting the zeros α and β.

Suppose all non-trivial zeros of grad $_\sigma(f)$ are **non-degenerate**. This means that $R^*(\Sigma)$ is **finite** and that the operators

$$- * d_A : T_{\mathcal{B}^*(P)}([A]) \to T_{\mathcal{B}^*(P)}([A])$$

which can be interpreted as the **Hessian** of f at a critical point $[A]$ are isomorphisms for all $[A] \in R^*(\Sigma)$. Under this simplifying assumption the instanton homology of Σ is defined as follows [F]: There exists a **relative Morse index**

$$i : R^*(\Sigma) \to \mathbb{Z}/_{(8)},$$

such that $i(\alpha) - i(\beta)$ is the dimension of $\mathcal{M}_\Sigma^\sigma(\alpha, \beta)$ modulo 8. The moduli spaces $\mathcal{M}_\Sigma^\sigma(\alpha, \beta)$ have various components whose dimensions differ by a multiple of 8; they come with a proper free $\mathbb{R}$-action arising from the translational symmetry of $\Sigma \times \mathbb{R}$ and can be oriented in a canonical way.
Moreover, $\mathcal{M}_\Sigma^\sigma(\alpha, \beta)$ has only finitely many 1-dimensional components for every pair of representations $\alpha, \beta \in R^*(\Sigma)$ with $i(\alpha) - i(\beta) = 1$. Comparing the induced orientation of these components with the natural orientation coming from the $\mathbb{R}$-action and taking the sum over the corresponding signs defines an integer $< \partial\alpha, \beta >$ [F].
Let $R_*(\Sigma) = \bigoplus_{i \in \mathbb{Z}/_{(8)}} R_i(\Sigma)$ be the graded free $\mathbb{Z}$-module whose components

$R_i(\Sigma)$ are generated by the elements $\alpha \in R^*(\Sigma)$ with $i(\alpha) = i$. The operators $\partial : R_i(\Sigma) \to R_{i-1}(\Sigma)$ given by

$$\partial\alpha = \sum_{i(\alpha)-i(\beta)=1} < \partial\alpha, \beta > \beta$$

satisfy $\partial \circ \partial = 0$, and can therefore be used to form the **instanton chain complex** $(R^*(\Sigma), \partial)$. The homology of this complex is a $\mathbb{Z}/_{(8)}$-graded abelian group $I_*(\Sigma) = \bigoplus_{i \in \mathbb{Z}/_{(8)}} I_i(\Sigma)$, which does not depend on the chosen metric σ. This is the **Floer** (or **instanton**) **homology** in the non-degenerate case. In general, i.e. if grad $_\sigma(f)$ has degenerate zeros, a suitable perturbation of the curvature equations has to be made in order to define $I_*(\Sigma)$ [F].

6.2 Seifert fibered homology spheres

Let $\pi : \Sigma \to \Sigma/_{S^1}$ be a Seifert fibration of a homology sphere Σ with n exceptional orbits $\pi^{-1}(x_i)$ of multiplicities $a_i, i = 1, \dots, n$. These multiplicities are necessarily pairwise relatively prime and the orbit space $\Sigma/_{S^1}$ is the 2-orbifold $(S^2; (x_1, a_1), \dots, (x_n, a_n))$. Conversely, given a set of n pairwise coprime integers $a_i \geq 2$ there exists a Seifert fibered homology sphere with these multiplicities; its diffeomorphism type is determined by $\underline{a} = (a_1, \dots, a_n)$ and will be denoted by $\Sigma(\underline{a})$ [NR].

The fundamental group of $\Sigma(\underline{a})$ has the following presentation:

Let $a := a_1 \cdot \ldots \cdot a_n$ and determine integers $b_i, i = 1, \dots, n$ by $0 < b_i < a_i$ and $b_i \frac{a}{a_i} \equiv -1 \pmod{a_i}$. Then

$$\pi_1(\Sigma(\underline{a})) = < t_1, \dots, t_n, h \mid h \,\text{central}\,, t_i^{a_i} = h^{b_i - a_i}, t_1 \cdot \ldots \cdot t_n = h^b >$$

for the unique integer b with $a(-b + \Sigma \frac{a_i - b_i}{a_i}) = 1$. This group is infinite with center $< h > \cong \mathbb{Z}$ if $n \geq 3$ except for $\pi_1(\Sigma(2,3,5)) \cong SL(2, \mathbb{F}_5)$ with center $< h > \cong \mathbb{Z}/_{(2)}$. The quotient $\pi_1(\Sigma(\underline{a}))/_{< h >}$ is the orbifold fundamental group $\pi_1^{\text{orb}}(\Sigma(\underline{a})/_{S^1})$ of the decomposition surface $(S^2; (x_1, a_1), \dots, (x_n, a_n))$ [FS2]. It is isomorphic to a cocompact Fuchsian group of genus 0 with the presentation

$$< t_1, \dots, t_n \mid t_i^{a_i} = 1, t_1 \cdot \ldots \cdot t_n = 1 > .$$

Fix a Seifert fibered homology sphere $\Sigma(\underline{a}) = \Sigma(a_1, \dots, a_n)$ with $n \geq 3$ exceptional fibers and consider a representation

$$\alpha : \pi_1(\Sigma(\underline{a})) \to SU(2).$$

If α is irreducible, then the generator h of the center must be mapped to the center $\{\pm 1\}$ of $SU(2)$; the images $\alpha(t_i)$ of the remaining generators are conjugate to diagonal matrices

$$\alpha(t_i) \sim \begin{pmatrix} \omega_i^{\ell_i} & \\ & \omega_i^{-\ell_i} \end{pmatrix}$$

with $\omega_i = \exp(\frac{2\pi\sqrt{-1}}{a_i})$ and certain numbers $\ell_i \in \mathbb{Z}/_{(a_i)}$. The invariants $(\pm\ell_1, \dots, \pm\ell_n)$ are the so-called **rotation numbers** of α [FS2].

Theorem 32 [FS2]. *The representation space $R^*(\Sigma(\underline{a}))$ of a Seifert fibered homology sphere is a closed differentiable manifold with several components. The rotation numbers of an irreducible representation α are invariants of the connected component of α in $R^*(\Sigma(\underline{a}))$. A component with rotation numbers $(\pm\ell_1,\ldots,\pm\ell_n)$ has dimension $2(m-3)$ where $m = \#\{i \mid 2\,\ell_i \neq 0\}$.*

Furthermore, there exists at most one component in $R^*(\Sigma(\underline{a}))$ realizing a given set of rotation numbers [BO]. In the special case of 3 exceptional orbits one obtains finite representation spaces whose points are non-degenerate critical points of the Chern-Simons function [FS2]. In the general case $n > 3$ these critical points are usually degenerate, so that the Chern-Simons function has to be perturbed in order to define the instanton homology of $\Sigma(\underline{a})$. R. Fintushel and R. Stern use a Morse function φ on the manifold $R^*(\Sigma(\underline{a}))$ to produce an appropriate perturbation. They show that the critical points of φ form a basis of the (perturbed) instanton chain complex. The relative Morse index of such a basis element α depends on the index of φ at α and on the rotation numbers of its component [FS2]. For any integer e let

$$R(\underline{a},e) = m - 3 + \frac{2e^2}{a} + \sum_{i=1}^{n} \frac{2}{a_i} \sum_{k=1}^{a_i-1} \cot\left(\frac{\pi a k}{a_i^2}\right) \cot\left(\frac{\pi k}{a_i}\right) \sin^2\left(\frac{\pi e k}{a_i}\right)$$

where $m = \#\{i \mid e \not\equiv 0 \,(\mathrm{mod}\, a_i)\}$.
This number is the virtual dimension of a certain instanton moduli space; it is always odd [FS2].

Theorem 33 [FS2]. *Let $\varphi : R^*(\Sigma(\underline{a})) \to \mathbb{R}$ be a Morse function on the representation space. The critical points of φ form a basis of the instanton chain complex. If α is a critical point of index $\mu_\varphi(\alpha)$ with rotation numbers $(\pm\ell_1,\ldots,\pm\ell_n)$, then its grading is given by*

$$i(\alpha) = -R(\underline{a},e) - 3 + \mu_\varphi(\alpha) \quad (\mathrm{mod}\, 8)$$

where the integer e satisfies $e \equiv \sum_{i=1}^{n} \ell_i \frac{a}{a_i} \,(\mathrm{mod}\, 2a)$.

In order to make explicit computations possible Fintushel and Stern describe $R^*(\Sigma(\underline{a}))$ as a configuration space of certain **linkages** in S^3 [FS2]. This method yields copies of S^2 as 2-dimensional components. On the basis of these examples they made the following

Conjecture 1 [FS2]. *The representation spaces $R^*(\Sigma(\underline{a}))$ admit Morse functions with only even index critical points.*

Note that this conjecture implies that the instanton chain complex is concentrated in even dimensions, so that the boundary operator has to be trivial. The Floer homology of $\Sigma(\underline{a})$ would then be torsion free, and could be read off from the rotation numbers and the Betti numbers of the components of $R^*(\Sigma(\underline{a}))$.

Theorem 34 [Bau], [BO], [FSt], [KK]. *The representation spaces $R^*(\Sigma(\underline{a}))$ admit Morse functions with only even index critical points. There exists a numerical algorithm which determines the Betti numbers of its components.*

There are at least three different proofs for the first part of this theorem [O]. The most direct proof is due to P. Kirk and E. Klassen [KK]. These authors use linkages in S^3 to construct the required Morse functions directly. The starting point for the other proofs as well as for the proof of the second part of the theorem is the following observation:
Every representation $\alpha : \pi_1(\Sigma(\underline{a})) \to SU(2)$ induces a representation $\bar{\alpha} : \pi_1^{\mathrm{orb}}(\Sigma(\underline{a})/_{S^1}) \to PU(2)$, such that the obvious diagram commutes:

$$\begin{array}{ccc} \pi_1(\Sigma(\underline{a})) & \xrightarrow{\alpha} & SU(2) \\ \downarrow /_{<h>} & & \downarrow /_{\mathbb{Z}/2} \\ \pi_1^{\mathrm{orb}}(\Sigma(\underline{a})/_{S^1}) & \xrightarrow{\bar{\alpha}} & PU(2). \end{array}$$

The mapping $\alpha \mapsto \bar{\alpha}$ identifies $R^*(\Sigma(\underline{a}))$ with the space

$$\mathrm{Hom}^*(\pi_1^{\mathrm{orb}}(\Sigma(\underline{a})/_{S^1}), PU(2))/_{\mathrm{conj.}}$$

of irreducible representations of the orbifold fundamental group in $PU(2)$. This representation space admits several different interpretations: It can be identified with a moduli space of **equivariant** Yang-Mills connections over a suitable covering of $\Sigma(\underline{a})/_{S^1}$. M. Furuta and B. Steer announced a proof of the Fintushel-Stern conjecture along these lines [FSt]. Extending the Atiyah-Bott method [AB] to this equivariant setting they also give formulas for the Poincaré polynomials of the moduli spaces.
The third approach is due to S. Bauer [Bau]. He interprets $\mathrm{Hom}^*(\pi_1^{\mathrm{orb}}(\Sigma(\underline{a})/_{S^1}), PU(2))/_{\mathrm{conj.}}$ as a moduli space of stable **parabolic** bundles over the marked Riemannian surface $(S^2; (x_1, a_1), \ldots, (x_n, a_n))$.
The algorithm for the computation of the Betti numbers has been found in the following way [BO]: Consider a rational elliptic surface over $\mathbb{P}^1$ — defined by a generic pencil of plane cubics — and perform logarithmic transformations with multiplicities $a_1, \ldots, a_n$ along smooth fibers over $x_1, \ldots, x_n \in \mathbb{P}^1$. The resulting elliptic surface $X(\underline{a}) = X(a_1, \ldots, a_n)$ is algebraic with fundamental group

$$\pi_1(X(\underline{a})) \cong \pi_1^{\mathrm{orb}}(\Sigma(\underline{a})/_{S^1}).$$

Choose a (sufficiently nice) very ample divisor $H(\underline{a})$ on $X(\underline{a})$ and let $\mathcal{M}_{X(\underline{a})}^{H(\underline{a})}(c_1, c_2)$ be the moduli space of stable vector bundles of rank 2 and Chern classes c_1, c_2 over $X(\underline{a})$. Denote the canonical class of $X(\underline{a})$ by $k(\underline{a})$. Then

$$\mathrm{Hom}^*(\pi_1^{\mathrm{orb}}(\Sigma(\underline{a})/_{S^1}), PU(2))/_{\mathrm{conj.}} \cong \mathcal{M}_{X(\underline{a})}^{H(\underline{a})}(0,0) \amalg \mathcal{M}_{X(\underline{a})}^{H(\underline{a})}(k(\underline{a}), 0)$$

as differentiable spaces [BO].
The moduli spaces $\mathcal{M}_{X(\underline{a})}^{H(\underline{a})}, (0,0)$ and $\mathcal{M}_{X(\underline{a})}^{H(\underline{a})}(k(\underline{a}), 0)$ can be described explicitly. They are **smooth complex projective** varieties with **rational** components. Every component comes with an algebraic **stratification** whose strata are Zariski open subsets in certain projective spaces. These subsets are complements of **secant varieties** of rational normal curves.

The normal bundles of the various strata are, however, hard to control, so that it becomes difficult to compute the Betti numbers topologically. The idea for avoiding this problem is to apply the **Weil conjectures**. An argument which is due to P. Deligne and L. Illusie [DI] allows to assume that the moduli spaces and their stratification are defined over an algebraic number ring. Then the Weil conjectures can be used to calculate the Betti numbers by counting points over finite fields.
The zeta functions of the strata are determined by the zeta functions of the secant varieties mentioned above. These functions are given by recursive formulas. The final result is an algorithm which computes the Betti numbers of $R^*(\Sigma(\underline{a}))$ from the initial data $\underline{a} = (a_1, \ldots, a_n)$; it can, in principle, be implemented on a computer [BO], [O].

Since the instanton homology $I_*(\Sigma(\underline{a}))$ of a Seifert fibered homology sphere is free over $\mathbb{Z}$ and vanishes in odd degrees, it is determined by the 4-tuple (r_0, r_1, r_2, r_3) of the ranks r_i of the groups $I_{2i}(\Sigma(\underline{a}))$.
For later use we list the Floer homology of a series of examples. The computations have been made by Fintushel and Stern [FS2].

Example 8.

$$I_*(\Sigma(2,3,6k+1)) = \begin{cases} (\frac{k-1}{2}, \frac{k+1}{2}, \frac{k-1}{2}, \frac{k+1}{2}) & \text{for } k \text{ odd} \\ (\frac{k}{2}, \frac{k}{2}, \frac{k}{2}, \frac{k}{2}) & \text{for } k \text{ even.} \end{cases}$$

7 Donaldson-Floer polynomials and singularities

Forthcoming work of Donaldson will define relative polynomial invariants for certain 4-manifolds W with homology sphere boundaries ∂W. These invariants are polynomials on $H_2(W;\mathbb{Z})$ with values in the Floer homology $I_*(\partial W)$ of the boundary. We describe these Donaldson-Floer polynomials for Milnor fibers of certain complete intersection surface singularities. The natural compactification in the weighted homogeneous case leads to strong non-vanishing results.

7.1 Relative invariants

The material in this section is based on Atiyah's expository article [A].
Let X be a simply-connected closed oriented topological 4-manifold with intersection form S_X. Suppose S_X has an algebraic decomposition $S_X = S^+ \oplus S^-$ as an orthogonal direct sum of forms S^+ and S^-. Then it follows from Freedman's classification theorem that such a decomposition admits a topological realization, i.e. there exist simply-connected closed oriented topological 4-manifolds X^+, X^- whose intersection forms S_{X^+}, S_{X^-} are equivalent to S^+, S^- respectively, so that X is homeomorphic to the connected sum $X^+ \# X^-$. The corresponding statement in the differentiable category is certainly not true, since e.g. the form $< E_8 >$ is not equivalent to the intersection form of a simply-connected smooth

4-manifold [Ma]. But even if the intersection form S_X of a smooth 4-manifold X splits as $S_X = S^+ \oplus S^-$, such that both S^+ and S^- are equivalent to intersection forms of closed differentiable 4-manifolds X^+ and X^-, it does not follow that X is diffeomorphic to the connected sum $X^+ \# X^-$; this is a consequence of Donaldson's indecomposability theorem for algebraic surfaces. The best one can hope for in the differentiable category is a **homology-connected-sum** decomposition.

Theorem 35 [FT]. *Let X be a simply-connected closed oriented differentiable 4-manifold whose intersection form splits as an orthogonal direct sum $S_X = S^+ \oplus S^-$. Then there exist simply-connected compact oriented differentiable 4-manifolds X^+, X^- with boundary $\partial X^+ = -\partial X^- = \Sigma$ a homology sphere and intersection forms S^+, S^- respectively, so that X is diffeomorphic to the homology-connected-sum $X^+ \cup_\Sigma X^-$.*

Suppose X is a simply-connected closed 4-manifold with a homology-connected-sum decomposition $X = X^+ \cup_\Sigma X^-$ such that $b_+(X^+) \neq 0$ and $b_+(X^-) \neq 0$. If the Donaldson invariants are defined for X, then there are two possibilities: either all polynomial invariants are trivial or the homology sphere Σ is distinct from the ordinary sphere S^3.
This observation suggests that the Donaldson polynomials factor in some sense through Σ [A].

In order to make this precise we need some notation.
The decomposition $X = X^+ \cup_\Sigma X^-$ and the corresponding direct sum decomposition $H_2(X; \mathbb{Z}) = H_2(X^+; \mathbb{Z}) \oplus H_2(X^-; \mathbb{Z})$ induces natural isomorphisms

$$S^d H_2(X; \mathbb{Z}) \cong \bigoplus_{r=0}^{d} S^r H_2(X^+; \mathbb{Z}) \otimes S^{d-r} H_2(X^-; \mathbb{Z}).$$

Let $I_*(\Sigma)$ be the Floer homology of Σ. There exists a natural **Poincaré duality** map

$$\bullet : I_*(\Sigma) \otimes I_*(-\Sigma) \to \mathbb{Z},$$

so that $I_j(-\Sigma) \cong I_{5-j}(\Sigma)$. Finally, let $\Phi_\ell(X)$ be a Donaldson polynomial associated to a principal $SU(2)$-bundle with second Chern class ℓ over X; $\Phi_\ell(X)$ is defined for $\ell > \frac{3}{4}(b_+(X) + 1)$ and is a homogeneous polynomial of degree $d(\ell) = 4\ell - \frac{3}{2}(b_+(X) + 1)$.

The following result of Donaldson has been announced by Atiyah [A].

Theorem 36. *Let W be a simply-connected compact oriented differentiable 4-manifold whose boundary is a homology sphere ∂W. There is a sequence of polynomials*

$$\varphi_r : S^r H_2(W;\mathbb{Z}) \to I_*(\partial W)$$

on $H_2(W;\mathbb{Z})$ with values in the Floer homology of ∂W. If X is a closed 4-manifold for which the $SU(2)$-invariants $\Phi_\ell(X)$ are defined, and if X has a homology-connected-sum decomposition $X = X^+ \cup_\Sigma X^-$ with $b_+(X^\pm) > 0$, then the following decomposition formula holds:

$$(*) \qquad \Phi_\ell(X) = \sum_{r=0}^{d(\ell)} \varphi_r(X^+) \bullet \varphi_{d(\ell)-r}(X^-).$$

In other words, the following diagramm commutes:

$$\begin{array}{ccc} S^{d(\ell)} H_2(H;\mathbb{Z}) & \xrightarrow{\Phi_\ell(X)} & \mathbb{Z} \\ \downarrow \cong & & \uparrow \bullet \\ \bigoplus_{r=0}^{d(\ell)} S^r H_2(X^+;\mathbb{Z}) \otimes S^{d(\ell)-r} H_2(X^-;\mathbb{Z}) & \xrightarrow{\sum_{r=0}^{d(\ell)} \varphi_r(X^+) \otimes \varphi_{d(\ell)-r}(X^-)} & I_*(\Sigma) \otimes I_*(-\Sigma). \end{array}$$

The construction of the Donaldson-Floer polynomials $\varphi_r(W)$ for 4-manifolds with boundary uses moduli spaces of instantons on W which extend a given flat connection on ∂W. The decomposition formula $(*)$ has two immediate consequences [A]: It proves the vanishing theorem for polynomial invariants of connected sums and it shows the non-triviality of some of the Donaldson-Floer polynomials for 4-manifolds occurring in homology-connected-sum decompositions of algebraic surfaces. Thus, in particular, the corresponding Floer homology groups must be non-trivial.
We will see in the next section that this includes all Seifert fibered homology spheres.

7.2 Surface singularities

Let $(Y, 0)$ be a 2-dimensional isolated complete intersection singularity whose link is a homology sphere [Lo]. A well-known example is the Brieskorn singularity with equation $z_1^2 + z_2^3 + z_3^5 = 0$; its link is the Poincaré sphere [Br]. Homology spheres which occur as links of isolated surface singularities have been classified in [EN], but it seems to be unknown which of them are links of complete intersections [NW].
Choose a suitable representative $\pi : \mathcal{Y} \to S$ of the semi-universal deformation of Y, and let $D \subset S$ be its discriminant [Lo]. Set $S' = S \backslash D, \mathcal{Y}' = \mathcal{Y} \backslash \pi^{-1}(D)$, and $\pi' = \pi|_{\mathcal{Y}'}$, so that $\pi' : \mathcal{Y}' \to S'$ is the projection of a locally trivial differentiable fiber bundle. The **Milnor fiber** of the singularity is the fiber $F = \pi^{-1}(s_0)$ for a

base point $s_0 \in S'$. F has the homotopy type of a bouquet of 2-spheres; its intersection form defines the **Milnor lattice** $L := H_2(F; \mathbb{Z})$ and the corresponding orthogonal group $O(L)$. There is a natural map $\psi : \mathrm{Diff}_+(F) \to O(L)$ from the component group of the group of orientation-preserving diffeomorphisms of F to $O(L)$. Let

$$\rho : \pi_1(S'; s_0) \to \mathrm{Diff}_+(F)$$

be the **monodromy representation** of the family $\pi' : \mathcal{Y}' \to S'$. The **monodromy group** Γ of the singularity $(Y, 0)$ is the image of $\psi \circ \rho$ in $O(L)$. It is generated by the **Picard-Lefschetz transformations** corresponding to the elements of a weakly distinguished basis of **vanishing cycles** [Lo]. The Picard-Lefschetz transformation associated to a vanishing cycle is the reflection in the orthogonal complement of this cycle and can be induced by a diffeomorphism of F which restricts to the identity on the link ∂F [Lo].
Let $V := H_2(F; \mathbb{C})$, let $-q$ be the intersection form of the Milnor fiber, and define Δ as the set of vanishing cycles. Then Δ satisfies the conditions (A), (B) and (C) of the algebraic theorem in section 4.4. If Σ is not the Poincaré sphere $\Sigma(2,3,5)$, then $b_+(F) > 0$, and hence Δ is an infinite set [EO1].

Theorem 37 [**EO1**]. *Let F be the Milnor fiber of a 2-dimensional isolated complete intersection singularity whose link is a homology sphere Σ distinct from the Poincaré sphere. If the Floer homology $I_*(\Sigma)$ is torsion free, then the Donaldson-Floer polynomials $\varphi_r(F)$ are trivial for r odd and of the form*

$$\varphi_r(F) = c_r q^{\frac{r}{2}}, \quad c_r \in I_*(\Sigma) \otimes \mathbb{Q}$$

for r even.

This follows from the invariance (up to sign) of the component functions of $\varphi_r(F)$ under the monodromy group of the singularity.

In order to obtain more explicit results we specialize to the Seifert fibered case. Every Seifert fibered homology sphere $\Sigma(\underline{a}) = \Sigma(a_1, \ldots, a_n)$ occurs as the link of a **Brieskorn complete intersection**

$$V_\Lambda(\underline{a}) = \{z \in \mathbb{C}^n \mid \sum_{j=1}^{n} \lambda_{ij} z_j^{a_j} = 0, i = 1, \ldots, n-2\},$$

where $\Lambda = ((\lambda_{ij}))$ is a generic $(n-2) \times n$ matrix of complex numbers [NR]. The Seifert fibration on the corresponding link $\Sigma(\underline{a}) = V_\Lambda(\underline{a}) \cap S^{2n-1}$ is induced by the natural S^1-action. Note that $I_*(\Sigma(\underline{a}))$ is torsion free.

The Milnor fiber of $V_\Lambda(\underline{a})$ has a **natural compactification** as a complete intersection in a weighted homogeneous space [EO1]. This compactification is a normal surface with finitely many quotient singularities. Let $X_\Lambda(\underline{a})$ be the minimal resolution of these singularities; $X_\Lambda(\underline{a})$ is a simply-connected [Do2] algebraic surface and has $p_g > 0$ unless $\Sigma(\underline{a})$ is the Poincaré sphere.

Theorem 38 [EO1]. *The Floer homology $I_*(\Sigma(\underline{a}))$ of a Seifert fibered homology sphere is always non-trivial. If F is the Milnor fiber of a Brieskorn complete intersection $V_A(\underline{a})$ with $\underline{a} \neq (2,3,5)$, then at least one of the Donaldson-Floer polynomials $\varphi_r(F)$ is non-trivial.*

The proof follows from Donaldson's non-vanishing theorem for the polynomial invariants Φ_ℓ of $X_A(\underline{a})$ and from his decomposition formula.

Let X be one of the surfaces $X_A(\underline{a})$. The divisor at infinity of X is a normal crossing divisor X_∞ whose components are smooth rational curves. A suitable linear combination of these curves is a canonical divisor K_X of X [EO1]. The intersection graph associated to X_∞ is a tree and can be determined by the following recipe [P]:
Let $\underline{a} = (a_1, \ldots, a_n), a := a_1 \cdot \ldots \cdot a_n$, and determine integers b_i with $0 < b_i < a_i$ and $b_i \frac{a}{a_i} \equiv -1 (\operatorname{mod} a_i)$. For each $i = 1, \ldots, n$ write $\frac{a_i}{a_i - b_i}$ as a continued fraction

$$\frac{a_i}{a_i - b_i} = [b_{i1}, \ldots, b_{ir_i}].$$

The intersection graph has then a central curve with weight $\frac{1}{a} - \sum_{i=1}^n \frac{a_i - b_i}{a_i}$ and n branches weighted by $-b_{ij}$.

Example 9. Consider the Brieskorn singularities

$$V(2,3,6k+1) = \{z \in \mathbb{C}^3 \mid z_1^2 + z_2^3 + z_3^{6k+1} = 0\}$$

with $k \geq 1$, and denote the minimal resolution of the compactification of its Milnor fiber by X^k. The dual graph of the curve at infinity is:

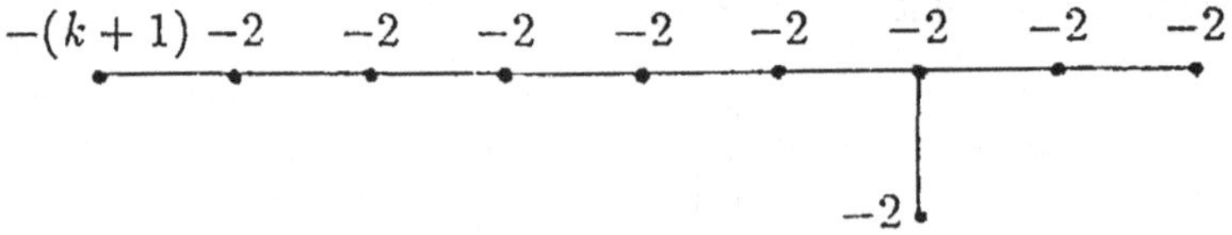

The curve X_∞ can be written as $X_\infty = C \cup Z$. Here Z is the union of the nine (-2)-curves which intersect according to the extended Coxeter-Dynkin diagram $\tilde{E}_8$, and C is a smooth rational curve with self-intersection number $-(k+1)$ intersecting Z transversally in one point. Let F_∞ be a singular elliptic curve of type II* supported on Z [BPV]. The associated linear system defines an elliptic fibration of X^k over $\mathbb{P}^1$ with F_∞ as fiber over ∞ [EO1].
The curve C is a section. The restriction of this fibration to the Milnor fiber is induced by the projection $(z_1, z_2, z_3) \mapsto z_3$; it has obviously $6k+1$ singular fibers of type II.

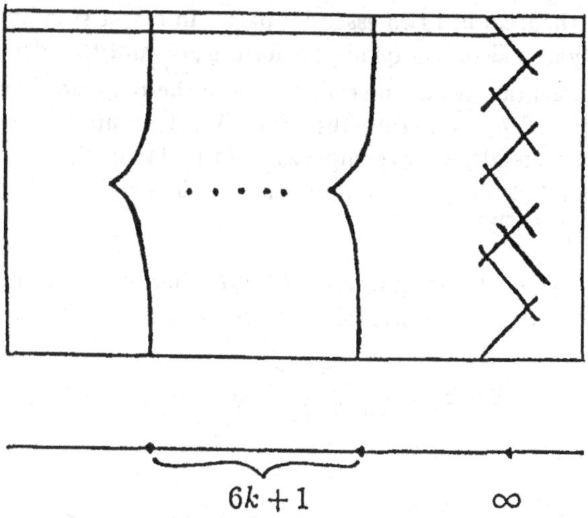

A canonical divisor of X^k is $K_{X^k} = (k-1)F_\infty$. Thus $p_g(X^k) = k$, and X^k is an elliptic $K3$ surface if $k = 1$ and has Kodaira dimension 1 if $k > 1$ [EO1].

Let X^{k-} be a tubular neighborhood of the curve $X_\infty \subset X^k$, and set $X^{k+} := X^k \setminus \mathrm{int}\,(X^{k-1})$, so that $X^{k+} \cap X^{k-} = \Sigma(2,3,6k+1)$. Perform logarithmic transformations with coprime multiplicities p and q along smooth fibers inside of X^{k-} and denote the resulting surfaces by $X^k_{p,q}$ and $X^{k-}_{p,q}$. The canonical class of $X^k_{p,q}$ has the form $(p\,q(k+1)-(p+q))\kappa_{X^k_{p,q}}$ with a primitive class $\kappa_{X^k_{p,q}}$. Let $q_{X^{k+}}$ and $q_{X^{k-}_{p,q}}$ be the quadratic forms of X^{k+} and $X^{k-}_{p,q}$ respectively.

Theorem 39 [EO1]. *The Donaldson-Floer polynomials $\varphi_r(X^{k+}) = c_r^+(k)q^{\frac{r}{2}}_{X^{k+}}$ have non-trivial coefficients for all even r. The Donaldson-Floer polynomials of $X^{k-}_{p,q}$ can be written in the form*

$$\varphi_r(X^{k-}_{p,q}) = \sum_{i=0}^{[\frac{r}{2}]} c^-_{ri}(k,p,q)\; q^i_{X^{k-}_{p,q}} \kappa^{r-2i}_{X^k_{p,q}}.$$

The coefficients satisfy relations

$$(*) \qquad \binom{N}{j} \frac{d!}{2^{N+1}N!}(pq)^k = c^+_{2j}(k)\cdot c^-_{d-2j,N-j}(k,p,q)$$

with $N = 2\ell - 2k - 1, d = 4\ell - 3(k+1)$ for every integer $\ell > 2(k+1)$ and all j with $0 \le j \le N$. The coefficients $c^-_{d-2j,N-j}(k,p,q)$ are contained in $I_*(-\Sigma(2,3,6k+1)) \otimes \mathbb{Q}$ and are non-trivial for all j with $0 \le j \le N$.

The proof of this theorem has two essential parts. In a first step one shows that $\varphi_r(X^{k-}_{p,q})$ is a polynomial in the quadratic form $q_{X^{k-}_{p,q}}$ and the class $\kappa_{X^k_{p,q}}$. This follows from the algebraic result in section 4 since the reflections corresponding to the components of F_∞ generate the affine Weyl group of type $\tilde{E}_8$ [EO1]. Then one applies Donaldson's decomposition formula to $X^k_{p,q} = X^{k+} \cup X^{k-}_{p,q}$ and uses Friedman and Morgan's computation of the leading coefficients of the polynomials $\Phi_\ell(X^k_{p,q})$ [FM4].

Example 10. Consider the singularity $V(2,3,7)$ and the corresponding compactification X. This is an elliptic $K3$ surface, so that the formulas simplify. One finds

$$\Phi_\ell(X) = \frac{(4\ell-6)!}{2^{2\ell-3}\cdot(2\ell-3)!}\; q_X^{2\ell-3}$$

for $\ell \geq 5$, and

$$\varphi_r(X^-) = c_r^- q_{X^-}^{\frac{r}{2}}, \quad c_r^- \in I_*(-\Sigma(2,3,7)) \otimes \mathbb{Q}.$$

The relations become

$$\binom{2\ell-3}{j}\frac{(4\ell-6)!}{2^{2\ell-3}\cdot(2\ell-3)!} = c_{2j}^+ \cdot c_{4\ell-6-2j}^-$$

for all $\ell \geq 5$ and all j with $0 \leq j \leq 2\ell-3$.

Note that the general form (*) of the relations shows that the exotic smooth structure of the surfaces $X^k_{p,q}$ is supported at infinity.
This phenomenon has been observed earlier by R. Gompf in a similar situation [Go]. He defines so-called **nuclei** of elliptic surfaces. It is not difficult to see that his nuclei are exactly the tubular neighborhoods of the curve at infinity for the singularities $V(2,3,6k+5)$ after the (three) exceptional curves have been contracted [EO1]. The special form of the Donaldson-Floer polynomials $\varphi_r(X^{k-}_{p,q})$ has consequences for the group of orientation-preserving diffeomorphism of $X^{k-}_{p,q}$ which restrict to the identity on the boundary, analogous to those in the closed case.

Up to now, none of the Donaldson-Floer polynomials has been determined explictly. The best result in this direction is due to Fintushel and Stern [FS3]: Let $X = X^+ \cup X^-$ be the $K3$ surface which compactifies the Milnor fiber of $V(2,3,7)$. Consider homology classes $\alpha_1,\alpha_2,\alpha_3,\alpha_4 \in H_2(X^-;\mathbb{Z}) = E_8 \oplus H$ with $\alpha_1,\alpha_2 \in E_8$, $\alpha_1^2 = \alpha_2^2 = -2$ and $\alpha_1\cdot\alpha_2 = -1$, $\alpha_3,\alpha_4 \in H$ with $\alpha_3^2 = \alpha_4^2 = 0$ and $\alpha_3\cdot\alpha_4 = -1$.
Similarly let $\alpha_5,\ldots,\alpha_{10} \in H_2(X^+;\mathbb{Z}) = E_8 \oplus H \oplus H$ be elements, so that α_5,α_6 form a pair in E_8, and α_7,α_8, and α_9,α_{10} form pairs in the two copies of H. For appropriate choices of such classes Fintushel and Stern show:

$$\begin{aligned}\varphi_4(X^-)(\alpha_1,\ldots,\alpha_4) &= m[\rho],\\ \varphi_6(X^+)(\alpha_5,\ldots,\alpha_{10}) &= n[\sigma],\end{aligned}$$

with odd integers m, n and generators $[\sigma] \in I_2(\Sigma(2,3,7)), [\rho] \in I_3(-\Sigma(2,3,7))$ such that $[\sigma] \cdot [\rho] = 1$. It follows that

$$c_4^- = \frac{m}{2^3}[\rho],\ c_6^+ = \frac{n}{2^4 \cdot 3}[\sigma].$$

It is very likely that algebraic techniques can be used to calculate Donaldson-Floer polynomials related to weighted homogeneous singularities.

8 References

[A] Atiyah, M.F.: New invariants of 3- and 4-dimensional manifolds. In: The Mathematical Heritage of Herman Weyl, Proc. Symp. Pure Math. **48**, 285 - 299 (1988)

[AB] Atiyah, M.F., Bott, R.: The Yang-Mills equations over Riemann surfaces. Phil. Trans. R. Soc. London A **308**, 523 - 615 (1982)

[AK] Altman, A.B., Kleiman, S.L.: Compactifying the Picard Scheme. Advances in Math. **35**, 50 - 112 (1980)

[AHS] Atiyah, M.F., Hitchin, N.J., Singer, I.M.: Self duality in four dimensional Riemannian geometry. Proc. Roy. Soc. London A **362**, 425 - 461 (1978)

[Ba] Barlow, R.: A simply connected surface of general type with $p_g = 0$. Invent. math. **79**, 293 - 301 (1985)

[B] Barth, W.: Moduli of vector bundles on the projective plane. Invent. math. **42**, 63 - 91 (1977)

[BPV] Barth, W., Peters, Ch., Van de Ven, A.: *Compact complex surfaces.* Erg. der Math. (3) 4. Berlin, Heidelberg, New York: Springer 1985

[Bau] Bauer, S.: Parabolic bundles, elliptic surfaces and $SU(2)$-representation spaces of genus zero Fuchsian groups. Preprint 67, MPI Bonn, 1989

[BO] Bauer, S., Okonek, C.: The algebraic geometry of representation spaces associated to Seifert fibered homology 3-spheres. Math. Ann. **286**, 45 - 76 (1990)

[Br] Brieskorn, E.: Rationale Singularitäten komplexer Flächen. Invent. math. **4**, 336 - 358 (1968)

[DI] Deligne, P., Illusie, L.: Relèvements modulo p^2 et décomposition du complex de de Rham. Invent. math. **89**, 247 - 270 (1987)

[Do1] Dolgachev, I.: Algebraic surfaces with $q = p_g = 0$. In: Algebraic Surfaces, C.I.M.E., p. 97 - 215 Liguori, Napoli (1981)

[Do2] Dolgachev, I.: Weighted projective varieties. In: Carrell, J.B. (ed.) Group actions and vector fields. Proc. Vancouver 1981. LNM **956**, 34 - 71, Berlin, Heidelberg, New York: Springer 1982

[D1] Donaldson, S.K.: Anti-self-dual Yang-Mills connections over complex algebraic surfaces and stable vector bundles. Proc. London Math. Soc. **50**, 1 - 26 (1985)

[D2] Donaldson, S.K.: Connections, cohomology and the intersection forms of 4-manifolds.
J. Diff. Geom. **24**, 275 - 341 (1986)

[D3] Donaldson, S.K.: Irrationality and the h-cobordism conjecture.
J. Diff. Geom. **26**, 141 - 168 (1987)

[D4] Donaldson, S.K.: Polynomial invariants for smooth 4-manifolds.
Topology, to appear

[D5] Donaldson, S.K.: The orientation of Yang-Mills moduli spaces and 4-manifold topology.
J. Diff. Geom., to appear

[D6] Donaldson, S.K.: Letter to the author, December 1988

[D7] Donaldson, S.K.: Talk at Durham, July 1989

[EO1] Ebeling, W., Okonek, C.: Donaldson invariants, monodromy groups, and singularities.
International Journal of Math., to appear

[EO2] Ebeling, W., Okonek, C.: On the diffeomorphism groups of certain algebraic surfaces.
Preprint, Bonn 1990

[EN] Eisenbud, D., Neumann, W.D.: *Three-dimensional link theory and invariants of plane curve singularities.*
Annals of Math. Studies 110, Princeton University Press, Princeton 1985

[EF] Elencwajg, G., Forster, O.: Vector bundles on manifolds without divisors and a theorem on deformations.
Ann. Inst. Fourier, Grenoble **32**, 4, 25 - 51 (1982)

[FS1] Fintushel, R., Stern, R.: $SO(3)$-connections and the topology of 4-manifolds.
J. Diff. Geom. **20**, 523 - 539 (1984)

[FS2] Fintushel, R., Stern, R.: Instanton homology of Seifert fibered homology 3-spheres.
Preprint, Univ. of Utah, Salt Lake City (1988)

[FS3] Fintushel, R., Stern, R.: Homotopy $K3$ surfaces containing $\Sigma(2,3,7)$.
Preprint, Univ. of Utah, Salt Lake City (1988)

[F] Floer, A.: An instanton invariant for 3-manifolds.
Commun. Math. Phys. **118**, 215 - 240 (1988)

[FU] Freed, D., Uhlenbeck, K.K.: *Instantons and four-manifolds.*
M.S.R.I. publ. no. 1. New York, Berlin, Heidelberg, Tokyo: Springer 1984

[Fr] Freedman, M.: The topology of 4-manifolds.
J. Diff. Geom. **17**, 357 - 454 (1982)

[FT] Freedman, M., Taylor, L.: Λ-splitting 4-manifolds.
Topology **16**, 181 - 184 (1977)

[FM1] Friedman, R., Morgan, J.W.: Algebraic surfaces and 4-manifolds: some conjectures and speculations.
Bull. A.M.S. Vol. 18, No. 1, 1 - 19 (1988)

[FM2] Friedman, R., Morgan, J.W.: On the diffeomorphism types of certain algebraic surfaces I.
J. Diff. Geom. **27**, 297 - 369 (1988)

[FM3] Friedman, R., Morgan, J.W.: On the diffeomorphism type of certain algebraic surfaces II.
J. Diff. Geom. **27**, 371 - 398 (1988)

[FM4] Friedman, R., Morgan, J.W.: Complex versus differentiable classification of algebraic surfaces.
Preprint, New York (1988)

[FMM] Friedman, R., Moishezon, B., Morgan, J.W.: On the C^∞ invariance of the canonical classes of certain algebraic surfaces.
Bull. A.M.S. Vol. 17, No. 2, 283 - 286 (1987)

[FuS] Fujiki, A., Schuhmacher, G.: The moduli space of Hermite-Einstein bundles on a compact Kähler manifold.
Proc. Japan Acad. **63**, 69 - 72 (1987)

[FSt] Furuta, M., Steer, B.: The moduli spaces of flat connections on certain 3-manifolds.
Preprint, Oxford (1989)

[G] Gieseker, D.: On the moduli of vector bundles on an algebraic surface.
Ann. of Math. **106**, 45 - 60 (1977)

[Go] Gompf, R.E.: Nuclei of elliptic surfaces.
Preprint, Univ. of Texas, Austin, 1989

[GH] Griffiths, P., Harris, J.: Residues and zero-cycles on algebraic varieties.
Ann. of Math. **108**, 461 - 505 (1978)

[H] Hempel, J.: *Three-Manifolds.*
Annals of Math. Studies 86, Princeton University Press, Princeton, 1967

[Hi] Hirzebruch, F.: *Topological methods in algebraic geometry.*
Grundlehren der math. Wiss. 131 (third edition). Berlin, Heidelberg, New York: Springer 1966

[HH] Hirzebruch, F., Hopf, H.: Felder von Flächenelementen in 4-dimensionalen Mannigfaltigkeiten.
Math. Ann. **136**, 156 - 172 (1956)

[I 1] Itoh, M.: On the moduli space of anti-self dual connections on Kähler surfaces.
Publ. R.I.M.S. Kyoto Univ. **19**, 15 - 32 (1983)

[I 2] Itoh, M.: The moduli space of Yang-Mills connections over a Kähler surface is a complex manifold.
Osaka J. Math. **22**, 845 - 862 (1985)

[Ki] Kim, H.J.: Moduli of Hermite-Einstein vector bundles.
Math. Z. **195**, 143 - 150 (1987)

[KK] Kirk, P.A., Klassen, E.P.: Representation spaces of Seifert fibered homology spheres.
Preprint, California Institute of Tech., Pasadena (1989)

[K1] Kobayashi, S.: First Chern class and holomorphic tensor fields.
Nagoya Math. J. **77**, 5 - 11 (1980)

[K2] Kobayashi, S.: *Differential geometry of complex vector bundles.*
Iwanami Shoten and Princton University Press (1987)

[Ko] Kodaira, K.: On homotopy $K3$ surfaces.
In: Essays on topology and related topics (ded. G. de Rham) 58 - 69. Berlin,

Heidelberg, New York: Springer 1970

[Kot] Kotschick, D.: On manifolds homeomorphic $CP^2\#8C\bar{P}^2$. Invent. math. **95**, 591 - 600 (1989)

[Ku] Kuranishi, M.: A new proof for the existence of locally complete families of complex structures. In: Aeppli, A., Calabi, E., Röhrl, H. (eds.): Proc. of the conference on complex analysis, Minneapolis 1964, 142 - 154, Berlin, Heidelberg, New York: Springer 1965

[La] Lamotke, K.: The topology of complex projective varieties after S. Lefschetz. Topology **20**, 15 - 51 (1981)

[L] Lawson, H.B.: *The theory of gauge fields in four dimensions.* Regional Conf. Series, A.M.S. 58. Providence, Rhode Island 1985

[Lo] Looijenga, E.J.N.: *Isolated singular points on complete intersections.* London Math. Soc. Lecture Note Series 77. Cambridge University Press, Cambridge 1984

[Lü1] Lübke, M.: Chernklassen von Hermite-Einstein Vektorbündeln. Math. Ann. **260**, 133 - 141 (1982)

[Lü2] Lübke, M.: Stability of Einstein-Hermitian vector bundles. Manuscr. Math. **42**, 245 - 257 (1983)

[LO] Lübke, M., Okonek, C.: Moduli spaces of simple bundles and Hermitian-Einstein connections. Math. Ann. **267**, 663 - 674 (1987)

[Ma] Mandelbaum, R.: Four-dimensional topology: an introduction. Bull. A.M.S., Vol. 2, No. 1, 1 - 159 (1980)

[Mar] Margerin, C.: Fibres stables et metriques d'Hermite-Einstein. Sém. Bourbaki, no. 683 (1987)

[M1] Maruyama, M.: Stable bundles on an algebraic surface. Nagoya Math. J. **58**, 25 - 68 (1975)

[M2] Maruyama, M.: Moduli of stable sheaves I. J. Math. Kyoto Univ. **17**, 91 - 126 (1977)

[M3] Maruyama, M.: Moduli of stable sheaves II. J. Math. Kyoto Univ. **18**, 557 - 614 (1978)

[M4] Maruyama, M.: Elementary transformations in the theory of algebraic vector bundles. In: Aroca, J.M., Buchweitz, R., Giusti, M., Merle, M. (ed.): Algebraic Geometry, Proc. La Ràbida, LNM 961, 241 - 266, Berlin, Heidelberg, New York: Springer 1982

[Mi] Miyajima, K.: Kuranishi family of vector bundles and algebraic description of the moduli space of Einstein-Hermitian connections. Publ. R.I.M.S. Kyoto Univ. **25**, 301 - 320, (1989)

[Mo1] Mong, K.-C. On some possible formulation of differential invariants for 4-manifolds. Preprint 34, MPI Bonn (1989)

[Mo2] Mong, K.-C.: Polynomial invariants for 4-manifolds of type $(1, n)$ and a calculation for $S^2 \times S^2$. Preprint 37, MPI Bonn (1989)

[Mu] Mukai, S.: Symplectic structure of the moduli space of sheaves on an abelian or $K3$ surface.
Invent. math. **77**, 101 - 116 (1984)

[NR] Neumann, W.D., Raymond, F.: Seifert manifolds, plumbing, μ-invariant, and orientation reversing maps.
In: Millet, K.C. (ed.) Algebraic and geometric topology, Proc. Santa Barbara 1977 LNM 664, 162 - 196, Berlin, Heidelberg, New York: Springer 1978

[NW] Neumann, W., Wahl, J.: Casson invariants of links of singularities.
Comment. Math. Helv. **65**, 58 - 78 (1990)

[O] Okonek, C.: On the Floer homology of Seifert fibered homology 3-spheres.
In: Proceedings, Durham 1989 (to appear)

[OSS] Okonek, C., Schneider, M., Spindler, H.: *Vector bundles over complex projective spaces.*
Progress in Math. 3. Boston, Basel, Stuttgart: Birkhäuser 1980

[OV1] Okonek, C., Van de Ven, A.: Stable bundles and differentiable structures on certain elliptic surfaces.
Invent. math. **86**, 357 - 370 (1986)

[OV2] Okonek, C., Van de Ven, A.: Γ-type-invariants associated to $PU(2)$-bundles and the differentiable structure of Barlow's surface.
Invent. math. **95**, 601 - 614 (1989)

[OV3] Okonek, C., Van de Ven, A.: Instantons, stable bundles and C^∞-structures on algebraic surfaces.
Preprint, Bonn (1989)

[P] Pinkham, H.: Normal surface singularities with $\mathbb{C}^*$-action.
Math. Ann. **227**, 183 - 193 (1977)

[Q] Quinn, F.: Isotopy of 4-manifolds.
J. Diff. Geom. **24**, 343 - 372 (1986)

[Sch] Schwarzenberger, R.L.E.: Vector bundles on algebraic surfaces.
Proc. London Math. Soc. (3) **11**, 601 - 623 (1961)

[S] Serre, J.-P.: *A course in arithmetic.*
Graduate Texts in Math. 7. New York, Heidelberg, Berlin: Springer 1973

[Sm] Smale, S.: An infinite dimensional version of Sard's theorem.
Amer. J. Math. **87**, 861 - 866 (1968)

[T] Taubes, C.H.: Self-dual connections on 4-manifolds with indefinite intersection matrix.
J. Diff. Geom. **19**, 517 - 560 (1984)

[Ue] Ue, M.: On the diffeomorphism types of elliptic surfaces with multiple fibers.
Invent. math. **84**, 633 - 643 (1986)

[U1] Uhlenbeck, K.K.: Removable singularities in Yang-Mills fields.
Commun. Math. Phys. **83**, 11 - 30 (1982)

[U2] Uhlenbeck, K.K.: Connections with L^p bounds on curvature.
Commun. Math. Phys. **83**, 31 - 42 (1982)

[UY] Uhlenbeck, K.K., Yau, S.-T.: On the existence of Hermitian-Yang-Mills connections in stable vector bundles.
Commun. Pure Appl. Math. **39**, 257 - 293 (1986)

[We] Wehler, J.: Moduli space and versal deformation of stable vector bundles. Rev. Roumaine Math. Pures Appl. **30**, 69 - 78 (1985)

[Y] Yau, S.T.: Calabi's conjecture and some new results in algebraic geometry. Proc. Nat. Acad. Sci. USA **74**, 1789 - 1799 (1977)

[Z] Zuo, K.: Generic smoothness of the moduli of rank two stable bundles over an algebraic surface.
Preprint 7, MPI Bonn, 1990

This article was processed using the LaTeX macro package with ICM style

C.I.M.E. Session on

Recent Developments in Geometric Topology and Related Topics

List of participants

G. ARCA, Dip. di Mat., Viale Merello, 09100 Cagliari

Chr. BAR, Math. Inst. d. Univ. Bonn, Meckenheimer Allee 160, 5300 Bonn 1

F. BARDELLI, Dip. di Mat., Corso Strada Nuova 65, 27100 Pavia

N. BOKAN, Otona Zupancica 34/28, 11070 Novi Beograd

D. BRANDT, Inst. de Math., Univ. de Génève, 2-4 rue du Lièvre, CH-1227 Acacias

E. CALABI, Dept. of Math., Univ. of Pennsylvania, Philadelphia, PA 19104-3859

B. COLBOIS, Dep. de Math. Appl., Ecole Polytechnique Fédérale, CH-1015 Lausanne

M. CORNALBA, Dip. di Mat., Corso Strada Nuova 65, 27100 Pavia

P. CRISTOFORI, Via Caneva 8, 44100 Ferrara

G. D'AMBRA, Dip. di Mat., Via Ospedale 72, 09100 Cagliari

M. D'APRILE, Dip. di Mat., Univ. della Calabria, 87036 Arcavacata di Rende (CS)

G. DE CECCO, Dip. di Mat., Via Provinciale Lecce-Arnesano, 73100 Lecce

S. DEMICHELIS, Dept. of Math., Harvard Univ., Science Center, One Oxford Str., 02138 Cambridge, (MA)

M. FALCITELLI, Dip. di Mat., Campus Universitario, Trav. 200 Re David 4, 70125 Bari

A. FARINOLA, Dip. di Mat., Campus Universitario, Via G. Fortunato, 70125 Bari

R. FERES, Math. Sci. Res. Inst., 1000 Centennial Drive, Berkeley, CA 94720

A. GASZAK, Inst. Math., A. Michiewicz Univ., Matejki 48/49, 60-769 Poznan

S. GOLDBERG, Dept. of Math., Univ. of Illinois, 273 Altgeld Hall, 1409 West Green St., Urbana, IL 61801

D. HUYBRECHTS, Humboldt-Universität zu Berlin, Mathematik GDR, P.O.Box 1297, DDR-Berlin 1086

J. KONDERAK, Ist. Mat., Univ. Salerno, 84100 Salerno

A. LANTERI, Dip. di Mat., Univ., Via C. Saldini 50, 20133 Milano

B. LEEB, Math. Inst. der Univ. Bonn, Beringstr. 3 / Zi.3, 5300 Bonn 1

M. MAMONE CAPRIA, Dip. di Mat., Univ., Via Vanvitelli 4, 06100 Perugia

V. MARINO, Dip. di Mat., Univ. della Calabria, 87036 Arcavacata di Rende (CS)

R.A. MARINOSCI, Dip. di Mat., Via Arnesano, C.P. 193, 73100 Lecce

P. MATZEU, Dip. di Mat., Univ., Via Opsedale 72, 09100 Cagliari

F. MERCURI, Departamento Matematica, UNICAMP, C.P. 6065, 13081 Campinas S.P.

M. MESCHIARI, Dip. di Mat., Via G. Campi 213/B, 41100 Modena

G. MESS, IHES, 35 Route de Chartres, 91440 Bures-sur-Yvette

C. MICHAUX

L. MIGLIORINI, Via Cimarosa 15, 50019 Sesto Fiorentino (FI)

P. MORASSI, Via Paluzza 21, 33028 Tolmezzo (VA)

E. MUSSO, Dip. di Mat., Viale Morgagni 67/A, 50134 Firenze

A. NANNICINI, Dip. di Mat. Appl., Fac. Ing., Via S. Marta 3, 50139 Firenze

M. PALLESCHI, Via Bergognone 27, 20144 Milano

R. PAOLETTI, Via F. Filzi 3, 50019 Sesto Fiorentino (FI)

A.M. PASTORE, Dip. di Mat., Campus Universitario, Trav. 200 Re David 4, 70125 Bari

D. PERRONE, Dip. di Mat., Via Arnesano, C.P. 193, 73100 Lecce

R. PIERGALLINI, Dip. di Mat., Via Vanvitelli, 06100 Perugia

G.P. PIROLA, Dip. di Mat., Strada Nuova 65, 27100 Pavia

M.P. PIU, Dip. di Mat., Via Ospedale 72, 09124 Cagliari

F. PODESTA', Scuola Normale Superiore, Piazza dei Cavalieri 7, 56100 Pisa

M. RENI, SISSA, Strada Costiera 11, 34014 Trieste

M. ROSSI, Dip. di Mat., Viale Morgagni 67/A, 50134 Firenze

M. SALVETTI, Scuola Normale Superiore, Piazza dei Cavalieri 7, 56100 Pisa

A. SANINI, Dip. di Mat., Pol. di Torino, Corso Duca degli Abruzzi 24, 10129 Torino

P. SITZIA, Via Stefano Cagna 2, 09126 Cagliari

A. SPIRO, Via delle Ginestre 3, 50100 Firenze

M. TROYANOV, Centre de Math., Ecole Polyt., F-91128 Palaiseau Cedex

S. VENTURINI, Dip. di Mat. U. Dini, Viale Morgagni 67/A, 50134 Firenze

M. WEBER, Math. Inst., Univ. Bonn, Wegelerstrasse 10, 5300 Bonn 1

G. WEILL, Dept. of Math., Polyt. Univ., 333 Jay Street, Brooklyn, NY 11201

FONDAZIONE C.I.M.E.
CENTRO INTERNAZIONALE MATEMATICO ESTIVO
INTERNATIONAL MATHEMATICAL SUMMER CENTER

"Continua with microstructures"

is the subject of the Third 1990 C.I.M.E. Session.

The Session, sponsored by the Consiglio Nazionale delle Ricerche and the Ministero della Pubblica Istruzione, will take place under the scientific direction of Prof. GIANFRANCO CAPRIZ (Università di Pisa) at Villa "La Querceta", Montecatini Terme (Pistoia), Italy, **from July 2 to July 10, 1990.**

Courses

a) **Invariants in the theory of crystal defects.** (6 lectures in English).
Prof. Cesare DAVINI (Università di Udine, Italy).

The course presents recent results on a continuum theory of defects in crystals. The theory is based on the notion that defects should be measured by descriptors which are additive over the parts of the crystal and which are invariant under elastic deformations. It is shown that there is an infinite list of descriptors with these properties with a finite functional basis. This complete list strictly includes the Burgers' vectors and the dislocation density tensor of the classical theory of dislocations. Connections with the basic mechanisms of crystal plasticity are also discussed.

Outline of the course

1. Crystal lattices. Old molecular theories of elasticity. Defects and their role in the mechanics of materials.
2. Continuous theories of defects: the contributions of Bilby and Kondo. A continuum model for defective crystals. Elastic invariants.
3. Characterization of the first order invariants and their interpretation. Invariants of higher order. Measures of local defectiveness.
4. Neutral deformations. Conjugacy and canonical states.
5. Characterization of canonical states. A complete list of invariants.
6. A connection with a classical theorem of Frobenius. Equidefective states. Slips and rearrangements.

Basic references

[1] Taylor, G.I., The mechanism of plastic deformation of crystals, Part I and II, Proc. Roy. Soc. A 145 (1934), 362-387, 388-404.
[2] Bilby, B.A., Continuous distributions of dislocations, In: "Progress in solids mechanics", Vol. I (I.N. Sneddon, ed.), North-Holland Publishing Co., Amsterdam, 1960.

[3] Kroner, E., Allgemeine Kontinuumstheorie der Versetzungen und Eigenspannungen, Arch. Rational Mech. Anal. 4 (1960), 273-334.
[4] Davini, C., A proposal for a continuum theory of defective crystals. Arch. Rational Mech. Anal. 96 (1986), 295-317.
[5] Davini, C., Elastic invariants in crystal theory, In: "Material instabilities in continuum mechanics and related mathematical problems" (J.M. Ball, ed.), Clarendon Press, Oxford, 1988.
[6] Davini, C. and B.P. Parry, On defect-preserving deformations in crystals. Int. J. Plasticity 5 (1989), 337-369.
[7] Davini, C. and G.P. Parry, A complete list of invariants for defective crystals, (to appear).

b) **Microstructural theories for granular materials.** (6 lectures in English).
Prof. James T. JENKINS (Cornell University).

Outline

We outline the derivation of continuum theories for granular materials that are appropriate in the two extremes of their behavior: rapid flows involving particle collisions, and quasi-static deformations with enduring, frictional, interparticle contacts. In each extreme the microstructural variable that is important is a symmetric second rank tensor. For rapid flows this tensor is the second moment of the velocity fluctuations. In quasi-static situations it is a measure of the orientational distribution of the contact area. In each case we discuss the determination of the microstructure from the appropriate field equations and assess its influence on the stresses necessary to maintain a given flow or deformation. For rapid flows, boundary conditions may be derived using methods similar to those employed to obtain the field equations and constitutive relations. We indicate how the boundary conditions influence flows and illustrate this by employing the results of the theory in a simple hydraulic model for a rock debris slide.

References

Jenkins, J.T., Cundall, P.A. and Ishibashi, I., Micromechanical modeling of granular materials with the assistance of experiments and numerical simulations, in "Powders and Grains" (J. Biarez and R. Gourves, eds.), pp. 257-264, Balkema: Rotterdam, 1989.
Jenkins, J.T., Balance laws and constitutive relations for rapid flows of granular materials, in "Constitutive Models of Deformation" (J. Chandra and R.P. Srivastav, eds.), pp. 109-119, SIAM, Philadelphia, 1987.
Jenkins, J.T., Rapid flows of granular materials, in "Non-classical Continuum Mechanics" (R.J. Knops and A.A. Lacey, eds.), pp. 213-225, University Press, Cambridge, 1987.

c) **Defects and textures in liquid crystals.** (6 lectures in English).
Prof. Maurice KLEMAN (Université Paris-Sud).

Outline

1. Microstructure, the director $\vec{n}$, equations of equilibrium.
2. Layered phases: the equation curl$\vec{n}$ = 0 and the geometry of focal conics; topology of defects at the Sm A - Sm C transition, analogy with monopoles.
3. Columnar phases; the equation div$\vec{n}$ = 0 and the geometry of developable domains.
4. Double helical patterns in sinectics and cholesterics; presence of minimal surfaces, frustration, a model for the chromosome of dinoflagellate.
5. Cubic phases and minimal surfaces.
6. Some related aspects in ferromagnets.

d) **The topological theory of defects in ordered media.** (6 lectures in English).
Prof. David MERMIN (Cornell University).

Outline

1. Example of ordered media and their associated spaces of internal states.

2. Defects and their physical importance, classes of mutually homotopic loops in the state space, and the relation between the two.
3. The fundamental group of the state space and its relation to the combination laws for defects; media with non-abelian fundamental groups.
4. Some simple topological properties of continuous groups; group theoretic characterization of the state space in terms of broken symmetry.
5. How to deduce the fundamental group directly from the symmetry of the uniform medium.
6. The second homotopy group, its relation to point defects in 3 dimensions, the conversion of point defects by moving them around line defects, and how to deduce all this directly from the symmetry of the uniform medium.

FONDAZIONE C.I.M.E.
CENTRO INTERNAZIONALE MATEMATICO ESTIVO
INTERNATIONAL MATHEMATICAL SUMMER CENTER

"Mathematical Modelling of Industrial Processes"

is the subject of the Fourth 1990 C.I.M.E. Session.

The Session, sponsored by the Consiglio Nazionale delle Ricerche and the Ministero della Pubblica Istruzione, will be under the auspices of ECMI (European Consortium for Mathematics in Industry) and in collaboration with SASIAM (School for Advanced Studies in Industrial and Applied Mathematics).

It will take place under the Scientific direction of Prof. VINCENZO CAPASSO (Director of SASIAM) and Prof. ANTONIO FASANO (Università di Firenze), in TECNOPOLIS (Valenzano, Bari), from **September 24 to September 29, 1990.**

Courses

a) **Case studies of Industrial Mathematics Projects.** (7 lectures in English).
Prof. Stavros BUSENBERG (Harvey Mudd College, Claremont, USA).

Outline

These lectures will describe several industrial projects in which I have been involved over the past twenty years. A number of these projects originated in the Claremont Mathematics Clinic Program where small teams of students and faculty study problems sponsored and funded by industrial concerns. The other projects originated in consulting activities or in different University-Industry mathematics programs.

The first lecture will give an overview of a variety of Industrial Mathematics problems and of the settings in which they arose. Each of the remaining lectures will be organized about a particular mathematical area which has been useful in specific projects. However, it is the nature of Industrial Mathematics that it cannot be easily encapsulated in tidy mathematical fields which are defined for the convenience of academics, and we will end up touching upon a variety of techniques and theories.

- Case Studies of Industrial Mathematics Problems
- Semiconductor Contact Resistivity: Inverse Elliptic Problems
- Inverse Problems: Examples, Theory, and Computation
- Adaptive Pattern Recognition via Neural Networks: Optimization
- Static and Dynamic Optimization Problems
- Agricultural and Animal Resource Management: Dynamical Systems.

General References

1. H.T. Banks and K. Kunisch, Estimation Techniques for Distributed Parameter Systems, Birkhauser, Boston, 1989.
2. D.P. Bertsekas, Constrained Optimization and Lagrange Multiplier Methods, Academic Press, New York, 1982.
3. Tarun Khanna, Foundations of Neural Networks, Addison-Wesley, Reading, Massachusetts, 1990.
4. C. Castillo-Chavez, S.A. Levin and C.A. Shoemaker (Eds.) - Mathematical Approaches to Problems in Resource Management and Epidemiology, Lecture Notes in Biomathematics 81, Springer Verlag, New York, 1989.

b) **Inverse Problems in Mathematics for Industry.** (7 lectures in English).
Prof. Bruno FORTE (University of Waterloo, Ontario, Canada).

Outline

The process of deriving a deterministic mathematical model from the knowledge of particular solutions(s) and/or global properties of solutions will be analyzed. Examples of inverse problems related to some typical industrial process will be presented. Mainly we will be dealing with: inverse problems in classical mechanics (dynamical systems), inverse problems in diffusion processes.

References

1. A.S. Galiullin, Inverse problems of dynamics, Mir Publisher, Moscow, 1984
2. Frederic Y.M. Wan, Mathematical models and their analysis, Harper and Row, New York, 1989.

c) **Mathematical Aspects of Some Industrial Problems.** (7 lectures in English).
Prof. Hendrik K. KUIKEN (Philips Research Lab., Eindhoven).

Lectures 1 and 2: The determination of surface tension and contact angle from the shape of a sessile drop.

Literature

C.A. Smolders and E.M. Duyvis, Contact angles and de-wetting of mercury. Recueil 80 (1961), 635-649.
C.J. Lyons, E. Elbing and I.R. Wilson, A general selected plane method for measuring interfacial tensions from the shape of pendant and sessile drops. J. Chem. Soc. Farad. Trans. 81 (1985), 327-339.
Y. Rotenberg, L. Buruvka and A.W. Newman, Determination of surface tension and contact angle from the shape of axisymmetric fluid interfaces. J. Coll. Interf. Sci. 93 (1983), 169-183.
H.K. Kuiken, The determination of surface tension and contact angle from shape of a sessile drop revisited. To be published.

Lecture 3 and 4: The mathematical modelling of viscous sintering processes

Literature

H.E. Exnor, Principles of single phase sintering. Revs. Powder Metall. Phys. Chem. 1 (1979), 7-251
H.K. Kuiken, Viscous sintering: the surface-tension-driven flow of a liquid form under the influence of curvature gradients at its surface. To appear in J. Fluid Mech.
H.K. Kuiken, Deforming surfaces and viscous sintering. To appear in Proc. Conf. on the Math. and Comp. of Deforming Surfaces. Cambridge 1988. Oxford U. Press 1990.

Lectures 5, 6 and 7: Mathematical modelling of etching processes

Literature

H.K. Kuiken, Etching: a two-dimensional mathematical approach. Proc. R. Soc. London A392 (1984), 199-225.
H.K. Kuiken, Etching through a slit. Proc. R. Soc. London A396 (1984), 95-117.
H.K. Kuiken, J.J. Kelly and P.H.L. Notten, Etching at resist edges. J. Elchem. Soc. 133 (1986), 1217-1226 (part 1), 1227-1232 (part 2).
H.K. Kuiken, Mathematical modelling of etching processes. Proc. 1987 Irsee Conf. on Free and Moving Boundaries. Pitman 1990.

LIST OF C.I.M.E. SEMINARS

		Publisher
1954 -	1. Analisi funzionale	C.I.M.E.
	2. Quadratura delle superficie e questioni connesse	"
	3. Equazioni differenziali non lineari	"
1955 -	4. Teorema di Riemann-Roch e questioni connesse	"
	5. Teoria dei numeri	"
	6. Topologia	"
	7. Teorie non linearizzate in elasticità, idrodinamica,aerodinamica	"
	8. Geometria proiettivo-differenziale	"
1956 -	9. Equazioni alle derivate parziali a caratteristiche reali	"
	10. Propagazione delle onde elettromagnetiche	"
	11. Teoria della funzioni di più variabili complesse e delle funzioni automorfe	"
1957 -	12. Geometria aritmetica e algebrica (2 vol.)	"
	13. Integrali singolari e questioni connesse	"
	14. Teoria della turbolenza (2 vol.)	"
1958 -	15. Vedute e problemi attuali in relatività generale	"
	16. Problemi di geometria differenziale in grande	"
	17. Il principio di minimo e le sue applicazioni alle equazioni funzionali	"
1959 -	18. Induzione e statistica	"
	19. Teoria algebrica dei meccanismi automatici (2 vol.)	"
	20. Gruppi, anelli di Lie e teoria della coomologia	"
1960 -	21. Sistemi dinamici e teoremi ergodici	"
	22. Forme differenziali e loro integrali	"
1961 -	23. Geometria del calcolo delle variazioni (2 vol.)	"
	24. Teoria delle distribuzioni	"
	25. Onde superficiali	"
1962 -	26. Topologia differenziale	"
	27. Autovalori e autosoluzioni	"
	28. Magnetofluidodinamica	"

1963 - 29. Equazioni differenziali astratte "
30. Funzioni e varietà complesse "
31. Proprietà di media e teoremi di confronto in Fisica Matematica "

1964 - 32. Relatività generale "
33. Dinamica dei gas rarefatti "
34. Alcune questioni di analisi numerica "
35. Equazioni differenziali non lineari "

1965 - 36. Non-linear continuum theories "
37. Some aspects of ring theory "
38. Mathematical optimization in economics "

1966 - 39. Calculus of variations Ed. Cremonese, Firenze
40. Economia matematica "
41. Classi caratteristiche e questioni connesse "
42. Some aspects of diffusion theory "

1967 - 43. Modern questions of celestial mechanics "
44. Numerical analysis of partial differential equations "
45. Geometry of homogeneous bounded domains "

1968 - 46. Controllability and observability "
47. Pseudo-differential operators "
48. Aspects of mathematical logic "

1969 - 49. Potential theory "
50. Non-linear continuum theories in mechanics and physics and their applications "
51. Questions of algebraic varieties "

1970 - 52. Relativistic fluid dynamics "
53. Theory of group representations and Fourier analysis "
54. Functional equations and inequalities "
55. Problems in non-linear analysis "

1971 - 56. Stereodynamics "
57. Constructive aspects of functional analysis (2 vol.) "
58. Categories and commutative algebra "

1972 -	59. Non-linear mechanics		"
	60. Finite geometric structures and their applications		"
	61. Geometric measure theory and minimal surfaces		"
1973 -	62. Complex analysis		"
	63. New variational techniques in mathematical physics		"
	64. Spectral analysis		"
1974 -	65. Stability problems		"
	66. Singularities of analytic spaces		"
	67. Eigenvalues of non linear problems		"
1975 -	68. Theoretical computer sciences		"
	69. Model theory and applications		"
	70. Differential operators and manifolds		"
1976 -	71. Statistical Mechanics		Ed Liguori, Napoli
	72. Hyperbolicity		"
	73. Differential topology		"
1977 -	74. Materials with memory		"
	75. Pseudodifferential operators with applications		"
	76. Algebraic surfaces		"
1978 -	77. Stochastic differential equations		"
	78. Dynamical systems		Ed Liguori, Napoli and Birhäuser Verlag
1979 -	79. Recursion theory and computational complexity		"
	80. Mathematics of biology		"
1980 -	81. Wave propagation		"
	82. Harmonic analysis and group representations		"
	83. Matroid theory and its applications		"
1981 -	84. Kinetic Theories and the Boltzmann Equation	(LNM 1048)	Springer-Verlag
	85. Algebraic Threefolds	(LNM 947)	"
	86. Nonlinear Filtering and Stochastic Control	(LNM 972)	"
1982 -	87. Invariant Theory	(LNM 996)	"
	88. Thermodynamics and Constitutive Equations	(LN Physics 228)	"
	89. Fluid Dynamics	(LNM 1047)	"

1983 -	90.	Complete Intersections	(LNM 1092)	"
	91.	Bifurcation Theory and Applications	(LNM 1057)	"
	92.	Numerical Methods in Fluid Dynamics	(LNM 1127)	"
1984 -	93.	Harmonic Mappings and Minimal Immersions	(LNM 1161)	"
	94.	Schrödinger Operators	(LNM 1159)	"
	95.	Buildings and the Geometry of Diagrams	(LNM 1181)	"
1985 -	96.	Probability and Analysis	(LNM 1206)	"
	97.	Some Problems in Nonlinear Diffusion	(LNM 1224)	"
	98.	Theory of Moduli	(LNM 1337)	"
1986 -	99.	Inverse Problems	(LNM 1225)	"
	100.	Mathematical Economics	(LNM 1330)	"
	101.	Combinatorial Optimization	(LNM 1403)	"
1987 -	102.	Relativistic Fluid Dynamics	(LNM 1385)	"
	103.	Topics in Calculus of Variations	(LNM 1365)	"
1988 -	104.	Logic and Computer Science	(LNM 1429)	"
	105.	Global Geometry and Mathematical Physics	(LNM 1451)	"
1989 -	106.	Methods of nonconvex analysis	(LNM 1446)	"
	107.	Microlocal Analysis and Applications	(LNM 1495)	"
1990 -	108.	Geoemtric Topology: Recent Developments	(LNM 1504)	
	109.	H_∞ Control Theory	(LNM 1496)	"
	110.	Continua with microstructures	to appear	"
	111.	Mathematical Modelling of Industrical Processes	to appear	"
1991 -	112.	Topological Methods in the Theory of Ordinary Differential Equations in Finite and Infinite Dimensions	to appear	